AF850828

LIPID METABOLISM and Its PATHOLOGY

LIPID METABOLISM and Its PATHOLOGY

Edited by
M. J. Halpern
Department of Biochemistry
and Center for Lipid Research
Faculty of Medical Sciences
Lisbon, Portugal

PLENUM PRESS • NEW YORK AND LONDON

Library of Congress Cataloging in Publication Data

International Colloquium on Lipid Metabolism and Its Pathology (1980: Lisbon, Portugal)
Lipid metabolism and its pathology.

"Proceedings of an International Colloquium on Lipid Metabolism and Its Pathology, held December 9-11, 1980, in Lisbon, Portugal"—T.p. verso.
Includes bibliographies and index.
1. Lipids—Metabolism—Disorders—Congresses. 2. Blood lipoproteins—Metabolism—Disorders—Congresses. I. Halpern, M. J. II. Title. [DNLM: 1. Cardiovascular Diseases—congresses. 2. Hyperlipoproteinemia—congresses. 3. Lipids—metabolism—congresses. 4. Lipoproteins—metabolism—congresses. 5. Metabolic Diseases—congresses. QU 85 I5916L 1980]
RC632.L5I58 1980 616.3'997 84-26445

ISBN-13: 978-1-4612-9486-3 e-ISBN-13: 978-1-4613-2445-4
DOI: 10.1007/978-1-4613-2445-4

Proceedings of an International Colloquium on Lipid Metabolism and Its Pathology, held December 9-11, 1980, in Lisbon, Portugal

©1985 Plenum Press, New York
Softcover reprint of the hardcover 1st edition 1985

A Division of Plenum Publishing Corporation
233 Spring Street, New York, N.Y. 10013

All rights reserved

No part of this book may be reproduced, stored in a retrieval system, or transmitted, in any form or by any means, electronic, mechanical, photocopying, microfilming, recording, or otherwise, without written permission from the Publisher

PREFACE

These colloquia were organized to stimulate the experience of scientists working in different fields of lipid metabolism and to update the state of the art of lipid and lipoprotein metabolism.

The main lectures and communications cover the lipid and lipoprotein structure, metabolism, epidemiology, laboratoral and clinical diagnosis, physiopathology, nutrition and pharmacology.

I hope that these proceedings will form a basis for dissemination of knowledge and will promote more research in these important areas.

M. J. Halpern

CONTENTS

METABOLIC AND CLINICAL SIGNIFICANCE OF THE LIPOPROTEIN FAMILY CONCEPT

Peter Alaupovic

Laboratory of Lipid and Lipoprotein Studies
Oklahoma Medical Research Foundation
Oklahoma City, Oklahoma 73104, USA

INTRODUCTION

Plasma lipoproteins represent a unique and complex system of macromolecular compounds, the principal function of which is to transport lipids throughout the organism and to deliver them to various cellular sites for utilization either as fuel or as structural components. Due to the presence of protein moieties (apolipoproteins) plasma lipoproteins are charged macromolecules and behave in the electric field as simple proteins. However, due to the presence of lipid constituents (neutral lipids, phospholipids and glycosphingolipids) plasma lipoproteins possess relatively low hydrated densities and behave in the gravitational field more as lipids than as simple proteins. The electrophoretic mobility and hydrated density have been utilized as the most important operational criteria for characterizing and classifying plasma lipoproteins. The application of zonal electrophoresis to plasma lipoproteins showed the presence of at least four major lipid-stained bands at the origin, and in the pre-β, β- and α_1-positions (1). Gofman and his coworkers (2) developed a preparative procedure for ultracentrifugal fractionation of plasma lipoproteins and observed a discontinuous distribution of lipoprotein particles along a density gradient ranging from 0.92 to 1.21 g/ml. On the basis of density and size distributions, plasma lipoproteins were classified into five major classes including chylomicrons, very low density (VLDL), low density (LDL), high density (HDL) and very high density (VHDL) lipoproteins. By early sixties, plasma lipoproteins were recognized as macromolecular distributions of lipid-protein complexes heterogeneous with respect to size and hydrated density. The operationally defined density classes or electrophoretic bands were accepted as the fundamental physical-chemical and metabolic entities of the system (3-5). Due to recognized significance of plasma lipids in the genesis and development of atherosclerosis, the emphasis was almost exclusively on the lipid constituents of plasma lipoproteins both from a clinical and metabolic point of view. Hyperlipoproteinemias were classified on the basis of electrophoretic or ultracentrifugal lipid patterns (5,6) and cholesterol and triglyceride were treated in most of the metabolic studies as if they were chemical entities independent of other lipoprotein constituents (7). Distribution analysis revealed a significant metabolic relationship between major lipoprotein density classes thought to result from and be related to the transport and metabolism of triglycerides (5). For example, the VLDL levels were found to be inversely related to the HDL levels; very high levels of VLDL were also inversely related to LDL_2. The available evidence indicated that VLDL are converted by a unidirectional lipolytic process to LDL (8, 9) and that this conversion may also result in the formation of some HDL particles (10). In contrast to the already available information regarding the composition, distribution and metabolism of the lipid constituents of lipoproteins, very little was known about the chemical nature and metabolic role of the protein moieties. It was generally accepted that LDL and HDL contain distinct protein constituents referred to as β- and α-protein, respectively, and that these apolipoproteins also constitute the protein complement of chylomicrons and VLDL (11, 12). The terminal amino acid analyses indicated, however, that triglyceride rich lipoproteins might contain apolipoproteins different from those characteristic of LDL and HDL (12, 13). To explore the possible presence of additional apolipoproteins, Gustafson et al. (14) submitted chylomicrons and VLDL from hypertriglyceridemic subjects to partial delipidization by n-heptane and isolated three distinct phospholipid-protein residues. Two of these phospholipid-protein residues were shown to contain the α- and β-protein, respectively, whereas the third phospholipid protein residue contained a protein characterized by serine and threonine as the N-terminal amino acids. Following Oncley's suggestion that the α-protein be called ApoA and β-protein ApoB, the third recognized apolipoprotein was called ApoC (14,15). In the meantime, the biochemical characterization of genetic disorders of lipid transport provided the most impressive evidence for the essential role of apolipoproteins in the formation of lipoprotein density classes (16). These two events, i.e., the discovery of ApoC and characterization of Tangier disease and abetalipoproteinemia as apolipoprotein deficiency diseases created a renewed interest in the chemistry and metabolic function of protein moieties and ushered the plasma lipoprotein field into its "apolipoprotein" phase of development. One of the more important results of this new trend was the discovery of additional minor apolipoproteins. One of these isolated from HDL and referred to initially as the "thin-line" polypeptide was designated ApoD (17), and another one isolated from VLDL and called initially the "arginine-rich" polypeptide was designated ApoE (18,19). This was followed by the isolation from HDL of an acidic polypeptide called ApoF (20) and another minor apolipoprotein called ApoG (21). Shore and Shore (22) showed that ApoA consists of two non-identical polypeptides called, according to the ABC nomenclature (23), A-I and A-II, while Brown et al. (24) demonstrated that ApoC is composed of three non-idential polypeptides referred to as C-I, C-II and C-III. Some apolipoproteins, such as ApoA-I, ApoC-III and ApoE, occur in several polymorphic forms (24-26). All these apolipoproteins were shown to be the integral components of plasma lipoproteins within the density range 0.92-1.21 g/ml. Studies on the quantitative determination of apolipoproteins have shown that ApoA and ApoB are the major and ApoC, ApoD, ApoE, ApoF and ApoG the minor apolipoproteins (20,21,27). However, in some dyslipoproteinemic states ApoC and ApoE may occur in concentrations equal to or greater than those of ApoA and ApoB (28). It is still not known whether apolipoproteins B, D, E, F and G consist of a single polypeptide or several non-identical polypeptides. There is some evidence to indicate that ApoB might consist of more than a single polypeptide (29-31) and/or that the intestinal ApoB differs structurally from the hepatic ApoB (32). The literature abounds with reports describing the occurrence of additional apolipoproteins such as the so-called ApoA-IV (33,34), the "threonine-poor" polypeptides (35), the "glycine-rich" polypeptide (36), the "proline-rich" polypeptide (37), the β_2-glycoprotein I or ApoH (38,39) and the "sugar-induced" polypeptides (40). However, insufficient information about their chemical and immunological characteristics prevents their recognition as integral components of the plasma lipoprotein system. Some of these polypeptides seem to be bound to lipids in the intestinal chylomicrons, but occur in plasma in the lipid free form. Others are only present in the plasma in response to a specific stimulus such as administration of antibiotics (35) or glucose infusion (40). It remains to be established whether these polypeptides should be considered as typical apolipoproteins capable of forming of and existing as plasma lipoproteins or whether they represent a special group of polypeptides only peripherally connected with the lipid transport system.

Immunologic characterization and quantitative analyses of apolipoproteins in major lipoprotein density classes showed a wide distribution of most, if not all, of the apolipoproteins throughout the entire density spectrum (28,41). This holds true for lipoproteins isolated from normolipidemic subjects and from patients with various types of dyslipoproteinemias (28).

An important question raised by the newly discovered apolipoprotein heterogeneity of major lipoprotein density classes relates to the mode of apolipoprotein localization on individual lipoprotein particles irrespective of their density characteristics. Are all apolipoproteins present on the same lipoprotein particle or does each apolipoprotein reside on a separate lipoprotein particle? To establish the localization of apolipoproteins in individual lipoprotein particles, we have studied the whole plasma and major lipoprotein density classes by double diffusion analyses using various combinations of monospecific antisera to all well characterized apolipoproteins (23). The interpretation of immunodiffusion patterns was based on Ouchterlony's principles (42) whereby a non-identity reaction indicates that any two apolipoproteins in question reside on separate lipoprotein particles, while an identity reaction suggests the presence of such apolipoproteins on the same lipoprotein particle; partial identity between two apolipoproteins denotes the presence of at least two separate lipoprotein particles, one of which contains both and the other a single apolipoprotein. Results from this and other laboratories (43-50) have shown that each density class contains several distinct lipoprotein particles rather than single homogeneous complexes. The finding that apolipoproteins in each density class occur in non-equimolar ratios provided further evidence that not all individual lipoprotein particles of the same density class could have the same apolipoprotein composition. Use of a procedure combining Sepharose 6B column chromatography and rocket electrophoresis permitted the size separation of plasma lipoproteins and quantitative determination of apolipoprotein distributions (51). The results showed that in normolipidemic subjects apolipoproteins A-I, A-II and B are distributed monomodally and apolipoproteins C and E are distributed bimodally. These distribution patterns indicated clearly the presence of ApoC and ApoE in both VLDL and HDL particles. Immunological testing of LDL and HDL revealed that these two density classes consist of discrete lipoprotein particles most of which are characterized by the presence of a single apolipoprotein (23,43,44,52). Similar studies with chylomicrons and VLDL showed that the triglyceride-rich lipoproteins also consist of several distinct lipoproteins, but that some of these lipoproteins contain two or more apolipoproteins (23,41). The most significant result of these studies has been the realization that each density segment contains a mixture of discrete lipoprotein particles which have similar hydrated densities but different apolipoprotein composition. This experimental findings are, thus, incompatible with the concept that operationally definded lipoproteins represent the fundamental chemical and metabolic entities of the lipid transport system.

To account for the lipoprotein heterogeneity of operationally defined lipoprotein classes, we have proposed that apolipoproteins be used as the specific and distinguishing markers for characterization and identification of lipoprotein particles (41,53). According to this conceptual view, lipoprotein particles which contain a single apolipoprotein are considered as the simplest physical-chemical entities of the lipid transport system and are referred to as lipoprotein families. A lipoprotein family is defined as a polydisperse system of lipoprotein particles characterized by the exclusive presence of a single apolipoprotein. An apolipoprotein is defined as a lipid-binding protein capable of forming a soluble, polydisperse lipoprotein family. Apolipoproteins may consist of a single polypeptide or may be composed of several non-identical polypeptides. To express the relationship between apolipoproteins and their corresponding lipoprotein families a new nomenclature was devised in which the apolipoproteins were designated by capital letters, the constitutive polypeptides by Roman numerals and the polymorphic forms of polypeptides by Arabic numerals (23,53). The lipoprotein families are named according to their corresponding apolipoproteins. The simplest lipoproteins forms of apolipoproteins A (A-I + A-II), B,. C (C-I + C-II + C-III), D, E, F and G are discrete lipoprotein families called LP-A, LP-B, LP-C, LP-D, LP-E, LP-F and LP-G. Discrete lipoprotein families tend to have hydrated densities greater than 1.030 g/ml and occur mainly in the LDL and HDL density segments. However, lipoprotein families also have tendency to interact to with one another to various degrees. For example, LP-C and LP-E interact avidly with the triglyceride-rich LP-B and triglyceride-rich LP-A to form association complexes with the hydrated densities and particle sizes characteristic of chylomicrons and VLDL (23,41). On the other hand, complexes formed between the triglyceride-poor LP-A and LP-D or LP-F (54) have high hydrated densities. The association complexes of lipoprotein families are named according to their apolipoprotein composition. A triglyceride-rich particle which contains ApoB, ApoC and ApoE is named LP-B:C:E. Another complex consisting of ApoA and ApoF is referred to as LP-A:F, etc. Osborne and Brewer (55) have suggested that the discrete, simple forms of lipoprotein families be referred to as "primary" lipoproteins and the associated forms of lipoprotein families as "secondary" lipoproteins. They also suggested that the equilibria between the "primary" and "secondary" lipoproteins are governed by the laws of mass action.

By recognizing apolipoproteins as the specific and unique constituents of lipoproteins, the concept of lipoprotein families has some practical and theoretical advantages over conceptual views based on non-specific properties such as particle size and hydrated density. The use of immunologic methodologies permits detailed studies on the formation, interactions and degradation of individual lipoprotein particles irrespective of their size, hydrated densities or lipid composition. Theoretically, the concept of lipoprotein families provides a framework which is flexible enough to accommodate any number of apolipoproteins or their corresponding lipoprotein particles into a chemically and metabolically integrated scheme of lipid transport.

The concept of lipoprotein families implies that the proper functioning of lipid transport depends on certain optimal concentrations of discrete and associated forms of lipoprotein families. Any perturbations of functional impairments of this physiological process should result in changed concentrations of lipoprotein families commensurate with the type and extent of the underlying metabolic defect. Therefore, each metabolic derangement ought to be characterized by a specific apolipoprotein profile or, more precisely, free and associated forms of lipoprotein families. To test this hypothesis, we have developed electroimmunoassays for apolipoproteins A-I, A-II, B, C-I, C-II, C-III, D, E and F and applied them to determination of apolipoprotein profiles in various dyslipoproteinemic states (28).

Serum lipids and apolipoprotein levels of normolipidemic women and men from the Oklahoma City area are presented in Tables I and II. Although this population sample is still too small for an adequate statistical evaluation,certain age-related trends in the concentration

Table I. Serum Apolipoprotein concentration of normolipidemic women

Age Group	Total Cholesterol	Triglyceride	Apolipoproteins						
			A-I	A-II	B	C-II	C-III	D	E
Years			mg/100 ml						
Women 20-39 (n = 39)	169±30*	76±28	135±25	70±17	83±22	2.8±1	7.8±1.6	7.2±2	10.8±3.3
Women 40-59 (n = 13)	205±21	98±33	149±17	66± 8	102±18	3.5±1	9.0±2	7.4±1.6	13.9±5.8
Women > 60 (n = 12)	227±28	92±28	163±35	70±15	114±19	3.7±0.5	9.7±1.4	9.2±1.6	13.3±4.3

*Means ± S.D.

Table II. Serum apolipoprotein concentrations of normolipidemic men

Age Group	Total Cholesterol	Triglyceride	Apolipoproteins						
			A-I	A-II	B	C-II	C-III	D	E
Years			mg/100 ml						
Men 20-39 (n = 30)	173±29*	83±25	115±19	55± 8	98±26	2.6±0.6	7.7±2.5	5.3±1.2	10.2±3
Men 40-59 (n = 11)	221±28	104±28	130±17	67±16	121±25	2.8±0.7	8.5±0.9	8.0±1	12.2±5.3
Men ≥ 60 (n = 7)	221±28	105±37	143±25	63±17	121±26	3.6±1	8.8±2.7	10.1±0.4	11.2±1.3

*Means ± S.D.

of apolipoproteins are already emerging from this preliminary study. In both sexes, the levels of apolipoproteins A-I, B, C-III, D and E increase slightly with age up to the end of the fifth decade with a subsequent plateau or decline. The levels of apolipoproteins A-II and C-II seem to remain constant during the adulthood. Studies carried out with large groups of selected North American populations (56-58) showed similar age-related trends in the levels of serum cholesterol and triglyceride; both of these lipids increase slightly with age up to the middle of the fifth age decade and then decrease slightly. In comparison with men, women have higher levels of ApoA-I and ApoA-II and lower levels of ApoB at all ages; the differences in the levels of apolipoproteins C-II, C-III, D and E seem to be negligible.

Hypolipoproteinemias such as Tangier disease, abetalipoproteinemia, hypobetalipoproteinemia and deficiency of lecithin:cholesterol acyltransferase are characterized by distinct apolipoprotein profiles These apolipoprotein patterns have already been described (28). Briefly, patients with Tangier disease have negligible concentrations of apolipoproteins A-I and A-II, but the reduction of ApoA-II is not commensurate with that of ApoA-I. There are significant differences between individual Tangier patients with respect to the levels of ApoB, ApoC and ApoE. This heterogeneity among Tangier patients is most probably due to differences in the extent of a defective catabolism of triglyceride-rich lipoproteins. The hallmark of abetalipoproteinemia is the absence of ApoB. However, other apolipoproteins are also affected in this familial disorder. The apolipoproteins A-I, A-II, C-I, C-II, C-III and D are significantly reduced and the molar ratio C-I/C-III is reversed from that found in normals. On the other hand, the ApoE levels seem to be normal. The heterozygotes for familial hypobetalipoproteinemia have very low levels of ApoB, but other apolipoproteins are within the normal concentration ranges. The only exception is the molar ratio of C-I/C-III which is, similar to abetalipoproteinemia, higher than in normal subjects. The apolipoprotein pattern of homozygotes with this familial disease is indistinguishable from the of abetalipoproteinemic patients. Patients with lecithin: cholesterol acyltransferase deficiency have reduced levels of apolipoproteins A, B, C and E, but normal levels of ApoD.

Studies on the apolipoprotein profiling of hyperlipoproteinemic phenotypes have indicated the occurence of four characteristic patterns (27,28). The first of these patterns is characterized by decreased levels of apolipoproteins A-I, A-II and B (phenotype I). The hallmark of the second profile is a proportional increase in the levels of all three ApoC peptides and ApoE (phenotypes III and V). The third profile is characterized by a significant increase of ApoB (phenotypes IIa and IIb). The characteristic feature of the fourth profile is a disproportionate increase in the concentration of ApoC-III (phenotypes IV and IIb).

Preliminary results of our recent studies on the quantification of serum apolipoproteins in patients with genetic hyperlipoproteinemias are shown in Table III. Patients with familial hypertriglyceridemia, familial hypercholesterolemia and familial combined hyperlipidemia are characterized by distinct concentrations of apolipopro-

Table III. Serum Apolipoprotein Concentrations of patients with familial hyperlipoproteinemias

Genotypes	Apolipoproteins				Ratio
	A-I	B	C-III	E	B/C-III
Familial Hypertriglyceridemia (n = 6)	120 ± 17	121 ± 24	34 ± 7	23 ± 11	3.6 ± 1.2
Familial Hypercholesterolemia (n = 14)	117 ± 26	193 ± 51	10 ± 3	14 ± 7	20 ± 5
Familial Combined Hyperlipidemia (n = 4)	136 ± 7	153 ± 35	21 ± 9	30 ± 12	7.7 ± 0.5
Familial Dysbetalipoproteinemia (n = 18)	116 ± 23	136 ± 39	22 ± 9	30 ± 11	6.5 ± 1.7
Normal Controls (n = 52)	130 ± 24	103 ± 25	8 ± 2	11 ± 4	13

teins B, C-III and E and, therefore, by distinct ratios of B/C-III. The B/C-III ratio of normal controls is approximately 12-13. Patients with familial hypertriglyceridemia have a disproportionate increase in the ApoC-III levels (with respect to ApoC-I and ApoC-II), an increased level of ApoE and a slight elevation of ApoB; the B/C-III ratio is very low. Due to normal levels of ApoC-III and ApoE, but very high concentration of ApoB, patients with familial hypercholesterolemia are characterized by very high ratios of B/C-III and B/E. Patients with familial combined hyperlipidemia have increased levels of ApoB and ApoE, but normal or slightly increased levels of ApoC-III; the values for B/C-III ratio fall in between those of normal subjects and patients with familial hypertriglyceridemia. Patients with familial dysbetalipoproteinemia can also be differentiated from patients with hypertriglyceridemia or hypercholesterolemia on the basis of B/C-III ratio. However, they might be differentiated from patients with combined hyperlipidemia only on the basis of higher levels of all three ApoC peptides. The apolipoprotein profiles of patients with familial lipoprotein lipase deficiency and patients with ApoC-II deficiency are very similar. The levels of ApoA-I, ApoA-II and ApoB are significantly reduced, while the levels of ApoC and ApoE are increased. The main difference between these two familial disorders is the absence of ApoC-II in patients with the familial C-II deficiency.

We propose that the apolipoprotein profiles characteristic of familial hyperlipoproteinemias reflect the most common biochemical defects of lipid transport. The increased levels of ApoB in patients with familial hypercholesterolemia are due to increased concentration of LP-B,whereas the elevated levels of ApoC-III in patients with familial hypertriglyceridemia may be due to the presence of triglyceride-rich lipoprotein particles (LP-C-III) which contain ApoC-III as the sole apolopoprotein. It is not yet known whether LP-C-III is a normally occurring lipoprotein particle or a particle formed as a result of overproduction of triglyceride. These two apolipoprotein profiles may, thus, be characteristic markers of anabolic defects or, in case of ApoB, of the receptor deficiency. The catabolic defects seem to be expressed by two other apolipoprotein profiles. The reduced levels of apolipoproteins A-I, A-II and B with concomitant increases of apolipoproteins C and E reflect, most probably, impaired degradation of triglyceride-rich LP-A:B:C:E association complexes (chylomicrons) caused either by the deficiency of lipoprotein lipase or by the deficiency of ApoC-II. The proportional increase of ApoB, ApoC and ApoE levels may indicate increased concentrations of various triglyceride-rich LP-B:C:E association complexes (VLDL and/or LDL_1) accumulating as a result of impaired degradation and/or removal of these lipoprotein particles.

Results from various laboratories (28,59-63) have shown that patients with coronary artery disease and patients with secondary hyperlipoproteinemias may also have characteristic apolipoprotein profiles. The development of new procedures for the quantification of apolipoproteins in discrete and associated forms of lipoprotein families may provide further refinement of this new approach to characterizing and monitoring lipid transport processes in health and disease and permit its introduction and wider application in clinical studies and practice.

REFERENCES

1. CRC Handbook of Electrophoresis, Vol. I, Lipoproteins: Basic Principles and Concepts, L.A. Lewis and J.J. Opplt, eds., CRC Press, Inc., Boca Raton, Florida, 1980.
2. Gofman, J.W., Lindgren, F.T. and Elliot, H., Ultracentrifugal Studies of Lipoproteins of Human Serum, J. Biol. Chem. 179, 973-979, 1949.
3. Gofman, J.W., Glazier, F., Tamplin, A., Strisower, B. and DeLalla, O., Lipoproteins, Coronary Artery Disease, and Atherosclerosis, Physiol. Rev. 34, 589-607, 1954.
4. Fredrickson, D.S. and Gordon, R.S., Jr., Transport of Fatty Acids, Physiol. Rev.38, 585-630, 1958.
5. Nichols, A.V., Human Serum Lipoproteins and Their Interrelationships, Adv. Biol. Med. Phys.11, 109-158, 1967.
6. Fredrickson, D.S., Levy, R.I. and Lees, R.S., Fat Transport in Lipoproteins: An Integrated Approach to Mechanisms and Disorders, N. Engl.J. Med. 276, 32-44, 94-103, 148-156, 215-226, 273-281, 1967.
7. Scanu,A., Factors Affecting Lipoprotein Metabolism, Adv. Lipid Res. 3, 63-138, 1965.
8. Gitlin, D., Cornwall, D.G., Na'Casato, D., Oncley, J.L., Hughes, W.L. and Janeway, C.A., Studies on the Metabolism of plasma Proteins in the Nephrotic Syndrome. II. The Lipoproteins, J. Clin.Invest.37, 172-186, 1958.
9. Shore, B. and Shore, V., Some Physical and Chemical Properties of the Lipoproteins Produced by Lipolysis of Human Serum S_f 20-400 Lipoproteins by Post-Heparin Serum, J. Atheroscler. Res. 2, 104-114, 1962.
10. Furman, R.H., Howard, R.P. and Alaupovic, P., Effect of Chronic Heparin Administration on Serum Lipids, Lipoproteins, Nitrogen and Electrolyte Balance in Normal and Heparin-Responsive and Heparin-Unresponsive Hyperglyceridemic Subjects, Metabolism 11, 879-892, 1962.
11. Avigan, J., Redfield, R. and Steinberg, D., N-terminal Residues of Human Serum Lipoproteins, Biochim. Biophys. Acta 20, 557-558, 1956.
12. Shore, B., C- and N-terminal Amino Acids of Human Serum Lipoproteins, Arch. Biochem. 71, 1-10, 1957.
13. Rodbell, M., N-terminal Amino Acid and Lipid Composition of Lipoproteins from Chyle and Plasma, Science 127, 701, 1958.
14. Gustafson, A., Alaupovic, P. and Furman, R.H., Studies of the Composition and Structure of Serum Lipoproteins: Separation and Characterization of Phospholipid-Protein Residues Obtained by Partial Delipidization of Very Low Density Liproteins of Human Serum, Biochemistry 5 , 632-640, 1966.

15. Gustafson, A., Alaupovic, P. and Furman, R.H., Studies of the Composition and Structure of Serum Lipoproteins: Physical-Chemical Characterization of Phospholipid-Protein Residues Obtained from Very Low Density Human Serum Lipoproteins, Biochim. Biophys. Acta 84, 767-769, 1964.
16. Herbert, P.N., Gotto, A.M. and Fredrickson, D.S., Familial Lipoprotein Deficiency (Abetalipoproteinemia, Hypobetalipoproteinemia and Tangier Disease). In The Metabolic of Inherited Disease, Fourth Edition, J.B. Stanburry, J.B. Wyngaarden and D.S. Fredrickson, eds., McGraw-Hill Book Co., New York, pp. 544-588, 1978.
17. McConathy, W.J. and Alaupovic, P., Studies on the Isolation and Partial Characterization of Apolipoprotein D and Lipoprotein D of Human Plasma, Biochemistry 15, 515-520, 1976.
18. Shore, V.G. and Shore, B., Heterogeneity of Human Plasma Very Low Density Lipoproteins, Separation of Species Differing in Protein Components, Biochemistry 12, 502-507, 1973.
19. Shelburne, F.A. and Quarfordt, S.H., A New Apolipoprotein of Human Plasma Very Low Density Lipoproteins, J. Biol. Chem. 249, 1428-1433, 1974.
20. Olofsson, S.O., McConathy, W.J. and Alaupovic, P., Isolation and Partial Characterization of an Acidic Apolipoprotein (Apolipoprotein F) from Human High Density Lipoproteins, Biochemistry 17, 1032-1036, 1978.
21. Ayrault-Jarrier, M., Alix, J.F. and Polonovski, J., Une nouvelle protéine des lipoprotéines du sérum humain: isolement et caractérisation partielle d'une apolipoprotéine G, Biochimie 60, 65-70, 1978
22. Shore, B. and Shore, V., Heterogeneity in Protein Subunits of Human Serum High-Density Lipoproteins, Biochemistry 7, 2773-2777, 1968.
23. Alaupovic, P., The Concepts, Classification Systems, and Nomenclatures of Human Plasma Lipoproteins, In CRC Handbook of Electrophoresis, Vol. I, Lipoproteins: Basic Principles and Concepts, L.A. Lewis and J.J. Opplt, eds., CRC Press, Inc., Boca Raton, Florida. pp. 27-46, 1980.
24. Brown, W.V., Levy,R.I. and Fredrickson, D.S., Studies of the Proteins in Human Plasma Very Low Density Lipoproteins J. Biol. Chem. 244, 5687-5694, 1969.
25. Lux, S.E. and John, K.M., Further characterization of the Polymorphic Forms of a Human High Density Apolipoprotein, apoLP-Gln-I (apoA-I), Biochim, Biophys. Acta 278, 266-270, 1972.
26. Utermann, G.,Isolation and Partial Characterization of an Arginine-Rich Apolipoprotein from Human Plasma Very Low-Density Lipoproteins: Apolipoprotein E, Hoppe-Seyler's Z. Physiol. Chem. 356, 1113-1121, 1975.

27. Alaupovic, P., Curry, M.D. and McConathy, W.J., Quantitative Determination of Human Plasma Apolipoproteins by Electroimunoassays, In International Conference on Atherosclerosis, L.A. Carlson, R. Paoletti, C.R. Sirtori and G. Weber, eds., Raven Press, New York, pp. 109-115, 1978.
28. Alaupovic, P., Structure and Function of Plasma Lipoproteins with Particular Regard to Hyperlipoproteinemias and Atherosclerosis, Ann. Biol. Clin. 38, 83-93, 1980.
29. Huang, S.S. and Lee, D.M., A Novel Method for Converting Apolipoprotein B, the Major Protein Moiety of Human Plasma Low Density Lipoproteins, into a Water-Soluble Protein, Biochim. Biophys. Acta 577, 424-441, 1979.
30. Olofsson, S.O., Bostrom, K., Svanberg, U. and Bondjers, G., Isolation and Partial Characterization of a Polypeptide Belonging to Apolipoprotein B from Low-Density Lipoproteins of Human Plasma, Biochemistry 19, 1095 1064 , 1980.
31. Kane, J.P., Hardman, D.A. and Paulus, H.E., Heterogeneity of Apolipoprotein B: Isolation of a New Species from Human Chylomicrons, Proc. Natl. Acad. Sci. USA 77, 2465-2469, 1980.
32. Krishnaiah, K.V., Walker, L.F., Borensztajn, J., Schonfeld, G. and Getz, G.S., Apolipoprotein B Variant Derived from Rat Intestine, Proc. Natl. Acad. Sci. USA 77, 3806-3810, 1980.
33. Weisgraber, K.H., Bersot, T.P. and Mahley, R. W., Isolation and Characterization of an Apoprotein from the d<1.006 Lipoproteins of Human and Canine Lymph Homologous with the Rat A-IV Apoprotein, Biochem. Biophy. Res. Commun. 85, 287-292, 1978.
34. Utermann, G. and Beisiegel, U., Apolipoprotein A-IV: A Protein Occurring in Human Mesenteric Lymph Chylomicrons and Free in Plasma. Isolation and Quantification, Eur. J. Biochem. 99, 333-343, 1979.
35. Shore, V.G., Shore, B. and Lewis, S.B., Isolation and Characterization of Two Threonine-Poor Apolipoproteins of Human Plasma High Density Lipoproteins, Biochemistry 17, 2174-2179, 1978.
36. Olofsson, S.O., Fager, G. and Gustafson, A., Isolation and Partial Characterization of a Glycine-and Serine Rich Polypeptide from Human Serum High-Density Lipoproteins (HDL), Scand. J. Clin.Lab. Invest. 37, 749-755, 1977.
37. Sata, T., Havel, R.J., Kotite, L. and Kane, J.P., New Protein in Human Blood Plasma, Rich in Proline, with Lipid-Binding Properties, Proc. Natl. Acad. Sci. USA 73, 1063-1067, 1976.
38. Polz, E., Kostner, G.M. and Holasek, A., Studies on the Protein Composition of Human Serum Very Low Density Lipoproteins: Demontration of the β_2-Glycoprotein-I, Hoppe Seyler's Z. Physiol. Chem. 360, 1061-1067,1979.

39. Nakaya, Y., Schaefer, E.J. and Brewer, H.B., Jr., Activation of Human Post Heparin Lipoprotein Lipase by Apolipoprotein H (β_2-Glycoprotein I), Biochem. Biophys. Res. Commun. 95, 1168-1172, 1980.
40. Malmendier, C.L., Christophe, J. and Ameryckx, J.P., Separation and Partial Characterization of New Apoproteins from Human Plasma High Density Lipoproteins, Clin. Chim. Acta 99, 167-176, 1979.
41. Alaupovic, P., Lee, D.M. and McConathy, W.J., Studies on the Composition and Structure of Plasma Lipoproteins. Distribution of Lipoprotein Families in Major Density Classes of Normal Human Plasma Lipoproteins, Biochim. Biophys. Acta 260, 689-707.
42. Ouchterlony, O., Handbook of Immunodiffusion and Immunoelectrophoresis, Ann Arbor Science Publishers, Inc., Ann Arbor, Michigan, 1968.
43. Kostner, G. and Alaupovic, P., Studies of the Composition and Structure of Plasma Lipoproteins. Separation and Quantification of the Lipoprotein Families Occurring in the High Density Lipoproteins of Human Plasma, Biochemistry 11, 3419-3428, 1972.
44. Lee, D.M. and Alaupovic, P., Physicochemical Properties of Low-Density Lipoproteins of Normal Human Plasma. Evidence for the Occurrence of Lipoprotein B in Associated and Free Forms, Biochem. J. 137, 155-167, 1974.
45. Albers, J.J., Chen, C.H. and Aladjem, F., Human Serum Lipoproteins. Evidence for Three Classes of Lipoproteins in Sf 0-2, Biochemistry 11, 57-63, 1972.
46. Kostner, G., Isolation and Characterization of Lipoprotein B from High-Density Human Serum Lipoproteins, Biochem. J.130, 913-917, 1972.
47. Fellin, R., Agostini, B., Rost, W. and Seidel, D., Isolation and Analysis of Human Plasma Lipoproteins Accumulating Postprandially in an Intermediate Density Fraction (d 1.006-1.019 g/ml), Clin. Chim. Acta 54, 325-333, 1974.
48. Olofsson, S.O. and Gustafson, A., Degradation of High-Density Lipoproteins (HDL) In Vitro, Scand. J. Clin Lab. Invest. 33, Suppl. 137, 57-62, 1974.
49. Ayrault-Jarrier, M., Alix, J.F. and Polonovski, J., Présence et isolement de deux lipoprotéines immunologiquement apparentées aux ApoA-I dans le serum humain, Biochimie 62, 51-59, 1980.
50. Weisgraber, K.H. and Mahley, R.W., Subfractionation of Human High Density Lipoproteins by Heparin-Sepharose Affinity Chromatography, J. Lipid. Res. 21, 316-325, 1980.
51. Weech, P., McConathy, W.J. and Alaupovic, P., Semi-Micro Procedures for the Isolation of Lipoproteins from Serum Specified by Their Apolipoprotein Content, In Abstracts

of Oral Presentations (Abstract 445), V International Symposium on Atherosclerosis, Houston, Texas, 1979.
52. Suenram, A., McConathy, W.J. and Alaupovic, P., Evidence for the Lipoprotein Heterogeneity of Human Plasma High Density Lipoproteins Isolated by Three Different Procedures, Lipids 14, 505-510, 1979.
53. Alaupovic, P., Conceptual Development of the Classification Systems of Plasma Lipoproteins, Protides Biol. Fluids Proc. Colloq. 19, 9-19, 1972.
54. Koren, E., McConathy, W.J. and Alaupovic, P. Characterization of Apolipoprotein F-Containing Lipoproteins in Human Plasma, Federation Proc. 40, 347, 1981.
55. Osborne, J.C., Jr. and Brewer, H.B., Jr., The Plasma Lipoproteins, Adv. Protein Chem. 31, 253-337, 1977.
56. Lavietes, P.H., Albrink, M.J. and Man, E.B., Serum Lipids of Normal Subjects with Aging, Yale J. Biol. Med. 46, 134-141, 1973.
57. Lindgren, F.T., Adamson, G.L., Jenson, L.C. and Wood, P. D., Lipid and Lipoprotein Measurements in a Normal Adult American Population, Lipids 10, 750-756, 1975.
58. The Lipid Research Clinics Program Epidemiology Committee, Plasma Lipid Distributions in Selected North American Populations: The Lipid Research Clinics Program Prevalence Study, Circulation 60, 427-439, 1979.
59. Avogaro, P., Bittolo Bon, G., Cazzolato, G. and Quinci, G. B., Are Apolipoproteins Better Discriminators than Lipids for Atherosclerosis? Lancet 1, 901-903, 1979.
60. Kladetzky, R.G., Assmann, G., Walgenbach, S., Touchert, P. and Helb, H. D., Lipoprotein and Apoprotein Values in Coronary Angiography Patients, Artery 7, 191-205, 1980.
61. Sniderman, A., Shapiro, S., Marpole, D., Skinner, B., Teng, B. and Kwiterovich, P.O., Jr., Association of Coronary Atherosclerosis with Hyperapobetalipoproteinemia (Increased Protein but Normal Cholesterol Levels in Human Plasma Low Density (β) Lipoproteins), Proc. Natl. Acad. Sci. USA 77, 604-608, 1980.
62. Gabor, J., Spain, M. and Kalant, N., Composition of Serum Very-Low-Density and High-Density Lipoproteins in Diabetes, Clin. Chem. 26, 1261-1265, 1980.
63. Schonfeld, G., Binge, C., Miller, J.P., Kessler, G. and Santiago, J., Apolipoprotein B Levels and Altered Lipoprotein Composition in Diabetes, Diabetes 23, 827-834, 1974.

ACKNOWLEDGMENTS

Studies reported in this paper were supported in part by U.S. Public Health Service Grant HL-23181 and by the resources of the Oklahoma Medical Research Foundation. I wish to thank my colleagues Drs. Diana M. Lee, Walter J. McConathy, Jim L. Kelley, C. Alan Suenram, Philip Weech and Eugen Koren for their support and help in formulating the theory of lipoprotein families. I thank Mr. Jim

D. Fesmire for his help with the apolipoprotein profiling of dyslipoproteinemias and Mrs. Madeline Farmer for her help in the preparation of this manuscript.

LIPIDS AND PROTEINS OF LIPOPROTEINS IN HUMAN ATHEROSCLEROSIS

P. Avogaro, G. Bittolo-Bon, G. Cazzolato,
F. Belussi, and E. Pontoglio

Regional General Hospital, National Council for Atherosclerosis
Unit of Venice, Italy

INTRODUCTION

Few papers in recent years have been devoted to the study of the chemical composition of lipoproteins in human atherosclerosis (Lewis et al. 1974, Carlson and Ericsson 1975, Avogaro et al. 1977). Moreover some studies have ascertained relevant variations in the plasma levels of some of the main apolipoproteins, namely ApoA-I, apoA-II and apoB (Berg et al. 1976, Albers et al. 1976, Avogaro et al. 1978, Bradby at al. 1978). This paper offers new data in this field which have been obtained through a study performed in a large population of survivors of myocardial infarction.

MATERIAL AND METHODS

350 patients, all male, aged 40-60 years were included in the present study. All had suffered an acutemyocardial infarction at least 6 months previously. As controls, 310 subjects were selected from a group of people voluntarily attending our unit as part of a Public Health Survey. These subjects were males, aged 40-60 years. Socioeconomic status was similar in controls and in patients.The criteria of selection of both, survivors and controls, have been published elsewhere (Avogaro et al. 1978).

In every subject the main lipoprotein classes (VLDL, LDL, HDL) were separated by preparative ultracentrifugation using a Spinco--Beckman L2-65B according to the Manual of Lipid Operations (1974)

Abbreviations: M.I. (Survivors of myocardial infarction; C (controls(; TC (Total cholesterol); TG (total triglycerides);LDL (low density lipoproteins); HDL (high density lipoproteins); B (apolipoprotein B); A_1 (apolipoprotein A_1).

Cholesterol was assayed by the method of Watson (1960) and triglycerides by the procedure of Eggstein (1977).

The plasma levels of apolipoproteins B and A-I were recorded by electroimmunoassay according to the procedure outlined by Curry et al. (1976). The technical details have been given exhaustively elsewhere (Avogaro et al. 1978).

The data were subjected to statistical analysis by means of Student's t-test and x^2test. The analysis was first carried out on the data recorded in the two series: patients and controls were then subgrouped by lipid phenotype according to the instructions of the WHO Memorandum (1970).

RESULTS

Survivors showed values of LDL-C and apoB significantly higher than in controls; while values of HDL-C and apoA-I were significantly lower (Tab. 1). Plasma values of cholesterol and triglycerides were similar in survivors and in controls. Analysis of the ratios showed significantly higher values of ratio LDL-C/HDL-C in survivors; in the same subjects values of apoA-I/apoB ratio was significantly lower than in controls. When subjects of both series, M.I. and controls, having normal plasma cholesterol and triglycerides values were sub-grouped the same results have been obtained. In the sub--group characterized by phenotype IIA total cholesterol, triglycerides, LDL-C and LDL-C/HDL-C ratio do not discriminate between survivors and controls. In subjects with IIB phenotype LDL-C, apoB and LDL-C/HDL-C ratio are higher in survivors while values of HDL-C, apoA-I and apoA-I/apoB ratio are lower. In subjects with phenotype IV only the two apoproteins and the apoA-I/apoB ratio discriminate between survivors and controls. As far as the discriminating power of the various analyzed parameters is concerned (Tab. 2) the ratio apoA-I/apoB has the highest power with the highest x^2 value and a contribution to total x^2 above 50%. Soon after comes in decreasing order the ratio LDL-C/HDL-C, apoB, LDL-C, apo-I a,d HDL-C. Total cholesterol and triglycerides do not have discriminating power.

DISCUSSION

Lipids and/or lipoproteins seem to be a key factor in the atherogenic process (Stamler 1967, Gofman et al. 1954, Kannel et. al. 1871). Studies of series of patients with ischemic heart disease have revealed that a combined elevation of both VLDL- and LDL-cholesterol was the most frequent abnormality (Lewis et al. 1974, Carlson and Ericsson 1975, Avogaro et al. 1977). Recently it has also been stressed that coronary heart disease is inversely related to levels of HDL-cholesterol (Miller and Miller 1975). Other authors have claimed that the levels of HDL-C are more predictive than the levels of LDL-C (Castelli et al. 1977). Nevertheless, a large proportion of patients who suffer a major clinical event from atherosclerosis may present a completely negative lipidemic and/or lipoproteinemic pattern.

Table 1. Values of some chemical components (lipids and proteins) of lipoproteins in survivors of myocardial infarction (M.I.) and in Controls (C) with normal phenotype (NL) or affected by hyperlipaemic state (IIA, IIB, IV).

	Age y	TC mg/dl	TG mg/dl	LDL-C mg/dl	HDL-C mg/dl	B mg/dl	A-I mg/dl	LDLC/HDLC	A-I/B
CONTROLS (260)	52	224	140	146	50	117	123	3.09	1.13
	8	46	86	42	10	31	22	1.24	.30
M.I. (308)	53	241**	140	168**	45**	151**	108**	3.99**	.58**
	8	49	72	46	10	34	19	1.49	.22
C. NL (181)	52	205	104	133	52	104	126	2.65	1.58
	9	29	29	27	9	21	21	.75	.38
M.I. NL (173)	54	214*	104	148**	45**	133**	108**	3.45**	.84**
	8	32	31	31	10	26	21	1.20	.23
C. IIA (23)	50	295	113	213	58	152	132	3.76	.88
	5	25	25	42	8	18	16	1.07	.15
M.I. IIA (63)	52	297	119	226	47**	183**	111**	5.06**	.62**
	9	29	26	27	9	26	14	1.47	.12
C. IIB (30)	51	290	254	203	41	164	110	5.14	.69
	7	29	56	32	7	27	12	1.59	.12
M.I. IIB (33)	50	299	234	207	42	180*	103*	1.50	.15**
	7	30	45	27	9	32	17	1.50	.15
C. IV (26)	49	213	286	121	41	135	108	3.18	.82
	6	21	139	26	12	20	24	1.14	.22
M.I. IV (39)	52	222	256	134	39	151*	103	3.56	.70*
	7	29	78	31	7	26	15	1.14	.16

* = $P < 0.05$ ** = $P < 0.01$

Table 2. Values of M.I. patients and controls beyond or below the corresponding mean value for the examined population as a whole. The statistical significance was obtained through the X^4 test.

				% of MI subjects
TC	>	90-th %-ile	(290 mg/dl)	13.6
TG	>	,,	(244 mg/dl)	10.1
LDL-C	>	,,	(200 mg/dl)	24.0
HDL-C	<	10-th %-ile	(37 mg/dl)	19.8
ApoA-I	<	,,	(97 mg/dl)	27.3
ApoB	>	90-th %-ile	(160 mg/dl)	36.4
LDL-C/HDL-C	>	,,	(4.434)	33.4
ApoA-I/ApoB	<	10-th %-ile	(0.767)	53.8

Recently a large series of 101 survivors of myocardial infarction was studied before their allocation to a surgical trial (Moore et al. 1979). Of the subjects in the trial, 43% had normal plasma lipid levels. Moreover there was no correlation between the extent of angiographycally proven arterial disease and plasma cholesterol, triglycerides, LDL-, VLDL-, or HDL-cholesterol levels.

Other researchers in recent years have stressed the reduced levels of apoA-I (Berg et al. 1976, Albers et al. 1976, Avogaro et al. 1978, Bradby et al. 1978) in atherosclerotic patients. This behaviour of apolipoproteins was recorded even in normolipemic patients; it seemed therefore that apolipoproteins may be more relevant than lipids in discriminating between patients and controls (Avogaro et al. 1979).

The present study has been extended to two large series of subjects. The obtained data stress the relevant role of apolipoproteins in discriminating between survivors and controls. Even when cholesterol, triglycerides, LDL-C and HDL-C the ratio LDL-C/HDL-C 'this limited to the type IIA and IV) were of no help, apoB, apoA-I and their ratio differed significantly between the two groups. The discriminating power of apoproteins and especially of apoB has been further stressed by the percentage of survivors having values of apoB above the mean value of all examined subjects (M.I. and controls).

A lipid hypothesis for atherogenesis has been argued for a long time on both clinical and experimental evidence. Despite the enormous volume of data gathered on this topic, the hypothesis is still weak, especially because many patients present atherosclerotic clinical pictures without being hyperlipidemic, and on the other hand there are hyperlipidemic patients who have not suffered any cardiovascular event (Harlan et al. 1966). This situation may explain the huge wave of interest in HDL, which are now considered even more predictive than LDL for cardiovascular events (Castelli et al. 1977). LDL were dominant since the time of Gofman et al. (1954) but now a few milligrams of HDL are considered to have more meaning. A few clinical studies (Berg et al.1976, Albers et al. 1976, Bradby et al. 1978, Avogaro et al. 1978) have recently stressed a possible major role for protein. This is particularly true as apolipoproteins are effective even when lipids do not work at all, as in the series presented here. From this data it appears the apoA-I and especially apoB are by far the best discriminators of all the components of lipoproteins, lipid or protein.

SUMMARY

The plasma levels of some lipids (C, TG, LDL-C, HDL-C), of some apoproteins (A-I and B) and of two ratios (LDL-C/HDL-C and apoA-I/apoB) have been recorded in a series of 350 survivors of of myocardial infarction and in 310 controls. The two groups as a whole were analyzed as well as sub-groups of subjects divided according to their lipid phenotype. Moreover the discriminant power of the various analyzed parameters has been proved x^2 test using the mean values obtained in the examined subjects (M.I. and controls) as cut off point. The data support the view that plasma levels of apoB and apoA-I as well as values of their ratio are the best indicators of lipid derangement in survivors of myocardial infarction.

REFERENCES

ALBERS J.J., WAHL P.W., CABANA G.V., HAZZARD W.B., HOOVER J.J. (1976): Quantitation of apolipoprotein A-I of human plasma high density lipoproteins. Metabolism, 25, 633.

AVOGARO P., BITTOLO BON G., CAZZOLATO G., QUINCI G.B., BELUSSI F. (1978): Plasma levels of apolipoproteins A-I and apolipoprotein B in human atherosclerosis. Artery, 4, 385.

AVOGARO P., BITTOLO BON G., CAZZOLATO G., QUINCI G.B., SANSO A., SPARLA M., ZAGATTI G/C/. CATURELLI G. (1978): Variations in apolipoproteins B and A-I in the course of acute myocardial infarction. Europ. J. Clin. Invest., 8, 121.

Avogaro P., CAZZOLATO G., TARONI G., BELUSSI F. (1977): Chemical composition of ultracentrifugation fractions in different pattern of human atherosclerosis. Atherosclerosis, 26, 163.

AVOGARO P., BITTOLO BON G., CAZZOLATO G., QUINCI G.B. (1979): Are apolipoproteins better discriminators than lipids for atherosclerosis? Lancet, 1, 901.

BERG K., BORRESEN A.L., DAHLEN G. (1976): Serum high density lipoprotein and atherosclerotic heart disease. Lancet 1, 499.

BRADBY G.V.H., VALENTE A.J., WALTON K.W. (1978): Serum high density lipoproteins in peripheral vascular disease. Lancet, 2, 1217.

CARLSON L.A., ERICSSON M. (1975): Quantitative and qualitative serum lipoprotein analysis. Part 2. Studies in male survivors of myocardial infarction. Atherosclerosis, 21, 435.

CASTELLI W. P., DOYLE J.Y., GORDON T.F., HAMES C.G., HJORTLAND M.C., HYLLEY S.B., KAGAN A., ZUKEL W.J. (1977): HDL cholesterol and other lipids in coronary heart disease. The cooperative lipoprotein phenotyping study. Circulation, 55 767.

CURRY M.D., ALAUPOVIC P., SUENRAM C.A. (1976): Determination of apolipoprotein A and its constitutive A-I and A-II polypeptides by separate electroimmunoassay. Clin. Chem., 22, 215.

EGGSTEIN M. (1977): Eine neue Bestimmung der Neutralfette in Blutserum und Gewebe. Part 2. Zuverlassigkeit der Method, andere Neutralfettbestimmungen, Normalwerte fur Triglyceride und Glycerin in menschlichem Blut. Klin. Aschr., 44, 267.

GOFMAN J.W., DE LALLA O., FREEMAN N.K., LINDGREN F.T., NICHOLS A.V., STRISOWER E.H., TAMPLIN A.R. (1954): Serum lipoprotein transport system in health, metabolic disorders, atherosclerosis and coronary artery disease. Plasma, 2, 413.

HARLAN W. R., GRAHAM J.B., HARVEY ENTES E. (1966): Familial hypercholesterolemia. A genetic and metabolic study. Medicine (Baltimore), 45, 77.

KANNEL A.B., CASTELLI W.P., GORDON T., McNAMARA P.M. (1971): Serum cholesterol, lipoproteins and the risk of coronary heart disease. The Framingham Study. Ann. Int. Med., 74, 1.

LEWIS B.,CHAIT A., OAKLEY C.M.D., WOOTTON I.D.P., KRIKLER D.M., ONITIRI A., SIGURDSSON G., FEBRUARY A. (1974): Serum lipoprotein abnormalities in patients with ischemic heart disease. Comparison with a control population. Brit. Med. J., 9, 489.

MANUAL OF LIPID OPERATIONS (1974), Lipid research clinics program lipid and lipoprotein analysis. U.S. Dept. of Health Education and welfare, Publ. n. (NIH) 75, Vol.I, Bethesda, MD, p. 628.

MILLER C.J., MILLER N.E. (1975): Plasma high density lipoprotein concentration and development of ischemic heart disease. Lancet, 1, 16.

MOORE R.B., LONG J.M., MATTA J.P., J.P., AMPLATZ K., VARCO R.L., BUCHWALD H. (1979): Plasma lipoproteins and coronary arteriography in subjects in the program on the surgical control of hyperlipidemias. Preliminary report. Atherosclerosis, 32, 101.

STAMLER J. (1967): Lectures on preventive cardiology. (Eds.: R. Grune, W. Stratton), New york.

WATSON D. (1960): A simple method for the determination of serum cholesterol. Clin. Chim. Acta, 5, 637.

WORLD HEALTH ORGANIZATION (1970), Memorandum on classification of hyperlipidemias and hyperlipoproteinemias. Bull. Wld. Hlth. Org., 43, 891.

COMPARISON OF THE VALUES OF HDL-CHOLESTEROL AND OF APOPROTEIN A_1 AND B IN THE ASSESSMENT OF CORONARY RISK

M.J. Bugugnani*, H. Fouyé*, R. Haiat**, and P. Desoutter**

Biochemical* and Cardiological** Departments

Hospital of Saint Germain on laye, 78104-France

ABSTRACT

The plasmatic values of apoprotein A, (Apo A_1), Apo B and of the classical lipidic components (cholesterol, triglycerides, HDL and LDL-cholesterol) were studied on 30 patients with acute coronary insufficiency (ACI) and 50 patients on the first day of an acute myocardial infarction (AMI), and compared with those of a control group of 150 subjects. The apoproteins were measured by electro-immunodiffusion on agarose gel. The apo A_1 was quantified by reference to a pool of normal serum, and apo B by reference to a titrated serum. The normal values obtained were ($m \pm S$):HDL-C: 1.16 ± 0.33 mmol/l apo A_1 = 0.85 ± 0.18; Apo B: 0.73 ± 0.22 g/l: LDL-C=4.02 ± 1.03 mmol/m. For patients with ACI and AMI, the values of HDL-C(1.14 ± 0.33) and of apo A_1(0.84 ± 0.18)were not different from those of the control group. On the other hand the values of apo B and of LDL-C were significantly increased in ACI, being respectively: 0.90 ± 0.20 and 4.82 ± 1.24 ($p<0.001$) and in case of AMI, respectively:0.84 ± 0.24 and 4.41 ± 1.36 ($p<0.01$).

In conclusion, apo B seems to have a greater power of discrimination for cardiovascular risk than HDL-C and apo A_1.

The plasmatic values of total cholesterol and of triglycerides are a poor indication of the degree of tissue atherosclerosis. Progress in this field has been achieved by measurement of HDL-cholesterol. Still more recently, the work of Alaupovic (1). Avogaro(2) and Fruchart (4) has shown the value of measuring apoproteins in the assessment of cardiovascular risk. Two apoproteins seem to be particularly interesting: apoprotein A_1(Apo A_1), which is the principal proteic constituent of the antiatherogenic HDL and apoprotein B (apo B), the marker of the atherogenic lipoproteins VLDL, IDL and LDL.

The aim of this work is to show their discriminating power for coronary atherosclerosis, compared with the other constituents of the lipid balance: triglycerides, HDL-cholesterol (HDL-C).

These measurements were taken during the first 24 hours of hospitalization of 80 patients in cardiological wards (53 men aged 62±13 years, 27 women aged 73±11 years), showing acute coronary insufficiency (30 cases) or acute myocardial infarction (50 cases).

These were compared with the measurements for a control group of 150 persons presumed to be healthy and a normal lipemia.

The cholesterol and triglyceride measurements were carried out using enzymatic techniques. The HDL-C was measured according to Finley's technique (3). The amount of LDL-cholesterol (LDL-C) was calculated according to Friedewald's formula.

$$\text{LDL-C} = \text{total cholesterol} - \frac{\text{triglycerides}}{5} - \text{HDL-C}$$

Measurement of the apoprotein A_1 and B was carried out by electroimunodiffusion on rehydratable agarose gel*, simultaneously containing the anti apo A_1 and anti apo B antibodies. The apo A_1 was quantified by reference to a pool of normal serum and apo B by reference to titrated serum supplied by the manufacturer. Table I gives the results obtained. It is interesting to note that the rate of total cholesterol is not a discriminating element, since this is often normal in our patients. The same applies to the rates of HDL-C and apo A_1, which remain similar to those for the control group. On the other hand the rates of LDL-C and apo B are significantly increased in the case of acute coronary insufficiency and of acute myocardial infarction.

Figure 1 shows the histograms of frequency of the apo A_1 and apo B values and Apo A_1/Apo B ratio. It also clearly shows the overlapping of the Apo A_1/Apo B ratio between the control group and two groups of patients. It thus appears that apo B has a better discriminating value than apo A_1 or HDL-C, since this was the only rate to be frankly pathological in cases of acute coronary disorders. This conclusion is not surprising, since it has been found that atheromatous lesions (see the works of Hoff (5) contain almost exclusively LP-B, i.e. lipoproteins showing the presence of apoprotein B.

Insofar as their measurement could become current practice, the question remains: can apoproteins help in the prediction of large-scale cardiovascular risk? Our first results seem encouraging, at all events.

---

* Sebia apofilm

Table I: Values for the different plasmatic lipoprotein constituents in control subjects and patients with acute coronary insufficiency and during the first 24 hours of acute myocardial infarction.

	Reference group	Acute coronary insufficiency	Myocardial infarction
Cholesterol (mmol/l)	5,75 ± 1,20	6,50 ± 1,34 *	6,29 ± 1,52
Triglycerides (mmol/l)	1, 05± 0,56	1,64 ± 0,87 ***	1,24 ± 0,57
HDL-Cholesterol (mmol/l)	1,16 ± 0,33	1,14 ± 0,33	1,13 ± 0,31
LDL-Cholesterol (mmol/l)	4,02 ± 1,03	4,82 ± 1,24 ***	4,41 ± 1,36
Apo A_1	0,85 ± 0,18	0,84 ± 0,18	0,85 ± 0,15
Apo B_1 (g/l)	0,73 ± 0,22	0,90 ± 0,20 ***	0,84 ± 0,24 **
Apo A_1 / Apo B	1,21 ± 0,36	0,97 ± 0,30 ***	1,09 ± 0,31 *

* p <0,05 ** p < 0,01 *** p < 0,001

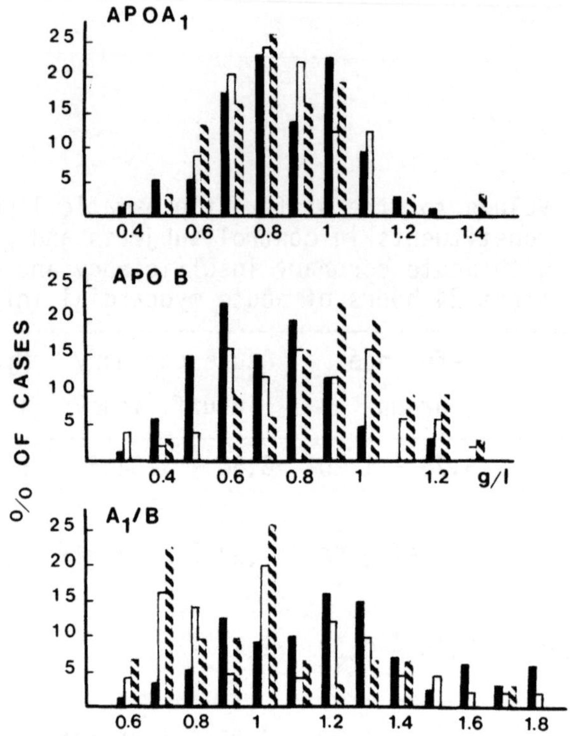

Figure I: Histogramas of frequency of plasmatic levels of Apo A_1, apo B and of apo A_1/apo B ratio, in reference group(■), acute myocardial infarction (□) and acute coronary insufficiency (▨).

REFERENCES

1. ALAUPOVIC P.: Structure and function of plasma lipoprotein with particular regard to hyperlipoproteinemias and atherosclerosis -Ann. Biol. Clin., 1980, 38, 83 - 93

2. AVOGARO P., BITTOLO BON G., QUINCI G.B. : Are apolipoproteins better discriminators than lipids for atherosclerosis - Lancet, 1979, 2, 901-903.

3. FINLEY P.R., SCHIFMAN R.B., WILLIAMS R., LICHTI D.A.: cholesterol in high density lipoprotein: use of Mg^{2+}/Dextran

Sulfate in its enzyme measurement. Clin Chem. 1978, 24, 931-933.
4. FRUCHART J. Ch., KORA I., CACHERA cl., CLAVEY V.-Simultaneous measurement of apolipoproteins A_1 and B by electroimmunoassay Clin. Chem. (sous presse).
5. HOFF H.F., HEIDEMAN C.L., GAUBATZ J.W., TITUS J.L., GOTTO A.M. Jr: Quantitation of Apo B in human aortic fatty streaks. A comparison with grossly normal intima and fibrous plaques. Atherosclerosis, 1978, 30, 263-272.

APOPROTEIN S VERSUS SAA PROTEIN

C.L. Malmendier and J.P. Ameryckx

Research Unit on Atherosclerosis, Laboratory of Chemical Pathology, Free University of Brussels Brussels, Belgium

ABBREVIATIONS AND NOMENCLATURE

VLDL: very low density lipoproteins (d < 1.006 g/ml)
HDL : high density lipoproteins (d=1.063-1.210 g/ml) and
HDL_2 and HDL_3: respectively d=1.063-1.125 and 1.125-1.210 g/ml

Apoprotein S_x: replaces $SV\text{-}D_x$ and means Subfraction x of DEAE cellulose chromatography of Fraction V obtained from Sephacryl S-200 chromatography of HDL apoproteins

PAGE: polyacrylamide gel electrophoresis
SAA : serum amyloid A protein

SUMMARY

Intravenous glucose infusions into neurological patients and post-surgery patients increased dramatically the amount of new polypeptides associated with high density lipoproteins, reaching up to 40% of total apoproteins whereas plasmas of normal subjects fed a regular diet contained only traces of these peptides (0.1%). This increase was observed in HDL_2 as well as in HDL_3 and also in very low density lipoproteins. Eight polymorphic forms of these new apoproteins named S (sugar induced) were isolated and defined by successive gel chromatography, ion-exchange chromatography, polyacrylamide gel electrophoresis and isoelectric focusing. Apoprotein S did not react with antisera prepared to apo B, apo C, apo A-I and apo A-II, but S_4 and S_5 gave a positive reaction with anti SAA.

Polypeptides S_4 and S_5 were recognized to have a high affinity for artificial lipid complexes. The change in the microviscosity of the apoprotein S-dimyristosyl-phosphatidylcholine and apo S-DMPC-cholesterol complexes with temperature, was similar to that observed for apo A-II and a stable complex might be isolated by sepharose 4B chromatography.

The simultaneous appearance of radioactivity into apoproteins S and other apoproteins of HDL (A-I, A-II and C) after the intravenous injection of 1-^{14}C-leucine suggests a common hepatic origin.

Aminoacid composition and partial sequence of apoprotein S were almost identical to those of SAA protein. Similarities and differences between apoprotein S and SAA protein are discussed. The putative precursor role played by apoprotein S in the formation of amyloid may build a link between glucose supply in the diet and tissue deposits observed in atherosclerosis.

INTRODUCTION

The name apoprotein S was recently assigned to a new family of apolipoproteins (1,2,3). The role in its production of glucose infusions administered to neurological patients was enforced by the 10-fold increase observed in a normal young subject only fed glucose (3) by intravenous route. It was however obvious that glucose was not the only factor responsible.

The present paper will deal with new facts about these S peptides and will try to sustain their belonging to the apoprotein family.

MATERIAL AND METHODS

Lipoproteins of the different density classes were isolated and purified by sequential ultracentrifugations. Lipoprotein preparations were exhaustively dialyzed and then delipidated in ethanol-ether. The polypeptides were separated and purified by gel permeation, and ion-exchange chromatography. Their purity was determined by amino acid analysis, polyacrylamide gel electrophoresis, and isoelectric focusing. Details about all methods were given in previous papers (1,2,3).

The protein content in lipoproteins and lipoprotein fractions was determined by a modification of the method of Lowry (4). Plasma cholesterol, HDL-cholesterol, triglycerides and phospholipids were determined using the procedures of the Lipid Research Clinics Program (5).

Apolipoproteins were tested by double diffusion and immunoelectrophoresis in 1 per cent agarose gels by using specific antisera to C-I, C-II, C-III, A-I, A-II and B. The immunological testing and the preparation of antisera were carried out as described by Schonfeld et al. (6,7). Apo S was also tested by immunodiffusion against anti SAA protein.

Phospholipid-cholesterol binding was evaluated by measuring fluorescence polarization and by the ability to isolate a stable lipid-protein complex by column chromatography (8).

In one patient the intravenous injection of 1-^{14}C-leucine was followed by the measurements of radioactivity incorporation into all proteins as a function of time. Specific activity curves were obtained in dpm/mg of protein or in dpm/μmole of leucine assuming a different number of moles of leucine per each apoprotein. The radioactivity of apolipoprotein bands after polyacrylamide gel electrophoresis was determined in gels stained with coomassie brilliant blue and sliced transversely with a razor blade giving slices 1-3 mm thick depending on the width of the protein bands. Each slice was dissolved in 2% periodic acid following the method of Anker (9), 10 ml of Instagel was added and the vials were counted in a Nuclear-Chicago Mark II liquid scintillation spectrometer.

RESULTS

a. Apoprotein composition

Plasma A-I and A-II levels in our normal subjects are similar to those described by Albers et al. (10) although systemically lower (Table 1). The difference is due to the mode of measurement of apoproteins. Our values correspond to the amount of apo A-I and apo A-II respectively in Fractions III and IV of Sephacryl S-200 chromatography. But Fractions I and II contain also apo A-I(about 10 mg/dl of plasma by electroimmunoassay) and the d>1.21 Fraction also contains 8 to 10 mg/dl of apo A-I (11).

Circa 50% of the weight of HDL are proteins and apo A-I and A-II (80% of HDL proteins) represent a valid index for alterations in HDL composition. Normally 43% of HDL_2 and 55% of HDL_3 (12) are proteins. Apo A-I/A-II ratio in our normal subjects is around 3.2 as compared to 2.7 (Ref. 11) to 3.6 (Ref. 10).

In all neurological and postsurgery patients both apo A-I and A-II levels are reduced to the half of normal values (Table 1) in such a way that the apo A-I/A-II ratio does not change very much, although being above normal. The percentage of the protein moiety in total HDL and in HDL_2 and HDL_3 does not differ from the normal values.

Apoprotein S replaces both apo A-I and A-II and decreased production of these apoproteins may be responsible for these changes. Absolute amounts of apoprotein S (Table 1) reach a mean value of 15.4 mg/dl in postsurgery patients and 31.5 mg/dl in neurological patients with extreme values up to 63 mg/dl whereas 0.5 mg/dl is the highest value observed in normal subjects.

Apoproteins C represent normally 7.6% of HDL proteins (13) or about 14 mg/dl. These values do not differ from those observed in our normal subjects (Table 3). There is a decrease of the apo C-II/C-III ratio in HDL of neurological patients. The postsurgery patients differ from normals by lower levels (in mg/dl) of all C proteins. Except on the 6th day of glucose infusion in the normal volunteer apoprotein C-II levels are not often markedly reduced (Table 3) which may explain the absence of any hypertriglyceridemia in all patients.

Figure 1 shows that marked changes in the HDL apoprotein pattern of one patient occur by modifying the nature of parenteral

Table 1. Concentration of Apoproteins A-I, A-II and S in the high density lipoproteins (mean values ± SD)

	HDL proteins mg/dl	A-I⁺ mg/dl	A-I⁺ %	A-II⁺ mg/dl	A-II⁺ %	A-I/A-II ratio	S⁺⁺ mg/dl
NORMAL SUBJECTS (5)	162 ± 7	100 ± 9	58.8 ± 2.7	31.5 ± 1.7	18.5 ± 1.4	3.2 ± 0.2	0.3 ± 0.15
NEUROLOGICAL PATIENTS (14)	114 ± 40	52.7 ± 20.7	43.6 ± 9.2	13.5 ± 5.1	11.0 ± 2.7	4.0 ± 1.0	31.5 ± 17.3
POSTSURGERY PATIENTS (9)	103 ± 17	41.5 ± 6.2	42.5 ± 6.1	11.3 ± 2.1	11.8 ± 3.4	3.8 ± 1.1	15.4 ± 7.0

\+ Protein content was determined in Fractions III and IV eluted from Sephacryl S-200 chromatography by the method of Schacterle and Pollack (ref.4) using bovine serum albumin as standard.

++ Apoprotein S represents the sum of 8 polymorphic forms recovered in Fraction V of Sephacryl S-200 chromatography and separated by PAGE excluding all apo C and D proteins.

Table 2. Chemical composition of Human Plasma high density lipoproteins

		MEAN VALUES			
		PROTEIN	TRIGLYCERIDE	CHOLESTEROL	PHOSPHOLIPID
		% BY WEIGHT			
NORMAL	HDL+	50	4	20	26
SUBJECTS	HDL_2+	43	4	23	29
	HDL_3+	55	4	19	22
NEUROLO-GICAL PATIENTS	HDL	49	6	17	28
	HDL_2	47	5	19	29
	HDL_3	57	3	12	28
POST-SURGERY PATIENTS	HDL	50	5	16	29
	HDL_2	43	8	20	29
	HDL_3	58	5	12	25

\+ HDL = high density lipoproteins d 1.063-1.210

HDL_2 d 1.063-1.125

HDL_3 d 1.125-1.210

Table 3. Comparison of Apoproteins C-I and C-II and C-III in high density lipoproteins between normal subjects and patients (Mean values ± SD)

		C-I		C-II		C-III			C-II/C-III
		mg/dl	%	mg/dl	%	mg/dl	%	%	Ratio
NORMAL SUBJECTS (5)		4.4 ± 0.6	3.5 ± 0.1	1.49 ± 0.29	1.18 ± 0.15	4.36 ± 1.00	3.42 ± 0.41	8.1	0.35 ± 0.07
NORMAL++ PERFUSED	3rd day	3.5	3.5	1.3	1.3	3.3	3.3	8.1	0.39
	6th day	3.5	4.1	0.75	0.65	1.7	1.5	5.7	0.43
NEUROLOGICAL PATIENTS (14)		3.0 ± 1.5	2.9 ± 0.7	1.30 ± 0.88	1.21 ± 0.56	5.10 ± 3.22	4.80 ± 2.28	8.9	0.26 ± 0.06
POSTSURGERY PATIENTS (9)		2.5 ± 0.4	3.2 ± 0.7	1.04 ± 0.24	1.35 ± 0.32	2.88 ± 0.66	3.57 ± 0.82	8.1	0.36 ± 0.11

\+ Quantitated by scanning of polyacrylamide gels of Sephacryl-Fraction V of HDL

++ Normal subject perfused intravenously with glucose for 6 days ; samples of the third and the sixth days

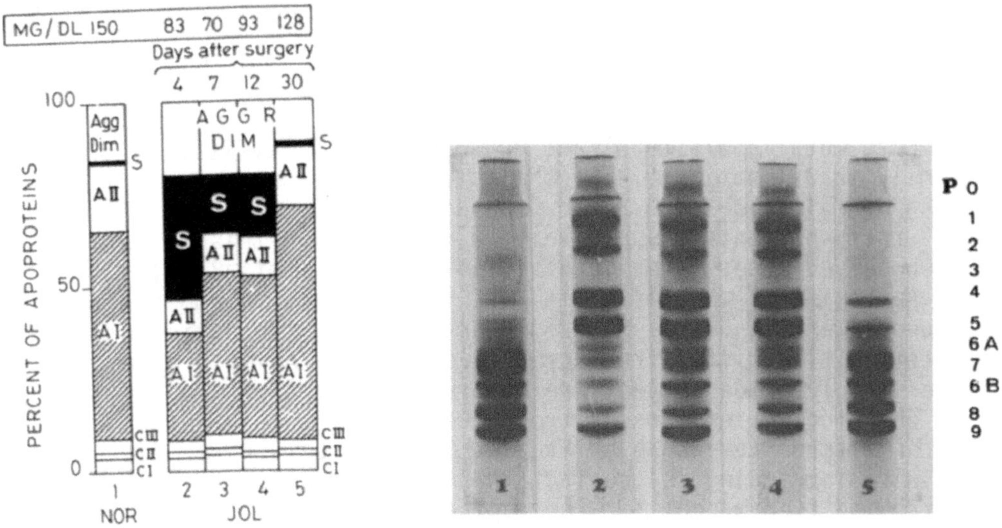

(A) (B)

Figure 1. Changes in apoprotein composition following modification in diet.

A. The first column on the left (NOR) shows the typical percentage pattern of HDL apoproteins in normal subjects. Columns 2 to 5 show the changes following substitution of intravenous perfusions of glucose for 4 days (2) by amino acid perfusions for 3 days (3), feeding by stomach tube for 5 days (4) and then after four weeks of a normal diet (5) in an hemicolectomized patient for cancer. The number of days after surgery are given above the columns. The corresponding protein values expressed in mg/dl of plasma and determined in HDL isolated by ultracentrifugation are enclosed by solid lines. Apoproteins A-1, A-II and C-I were recovered in Fractions III, IV and VI of Sephacryl S-200 chromatography of apo HDL respectively. Apo S, C-II and c-III were obtained by scanning of PAGE gels (Davis, 8M urea, 7% acrylamide, pH 8.6) from Fraction V of Sephacryl chromatography.

B. Polyacrylamide gel electrophoresis of Sephacryl Fraction V of apo-HDL (for conditions see 1 A). Tube 1 = normal patient.
Tubes 2 to 5 = as in 1 A

infusions and by restoring a normal diet. The percentage of apoprotein S, as high as 34 of the total at the beginning, diminishes to 16 when glucose infusions are replaced by amino acid infusions (Travasol 3.5%) for three days. It takes more than four weeks of a normal diet before normalization of both fractions apo A-I and A-II occurred. The increase in total HDL proteins follows more slowly.

b. Lipid composition

Both total cholesterol and HDL-cholesterol levels fall significantly in all patients. The cholesterol/protein ratio decreases from a normal value of 0.35 to 0.30 and 0.28 in neurological and postsurgery patients respectively. A reduction in this ratio was also observed after carbohydrate feeding by Blum et al. (14).

Approximately 50% of normal HDL is protein, 28% phospholipid, 21% cholesterol and only 3% triglyceride but in the patients, this HDL composition differ from normal, as a lower cholesterol and higher triglyceride content are observed (Table 2).

The glucose infusions although decreasing both HDL-A-I and HDL-cholesterol levels do not appear to increase VLDL production.

Also intriguing is the fact that HDL composition is not greatly modified except for the replacement of apoprotein A by apoprotein S. This implies that either remaining apo A-I, A-II and C of the subjects are binding much more lipid than normally to maintain a normal lipid/protein ratio or more likely that the substitutive apoprotein S is able to bind lipids in vivo almost as efficiently as do apoproteins A-I and A-II in normal conditions.

c. Immunological properties

Apoprotein S (forms S_2, S_4 and S_5) do not react with antisera to any other human plasma apolipoprotein tested (A-I, A-II, B, C-I, C-II, C-III). The proteins S_4 and S_5 however gave a positive reaction with antisera to human SAA on immunodiffusion (Fig. 2) but S_o, S_1 and S_2 gave no reaction of identity. As differences in a small number of residues may change the biological activity of a protein without affecting its immunological activity, it is difficult to conclude that both apo S and SAA protein have more than a similar behavior of immunologically active similar parts until the full sequence of all forms of apo S will be known.

d. Origin of apoprotein S

The intravenous administration of leucine-1-^{14}C into a neurological patient resulted in a substantial labeling of apo A-I and A-II. The specific activity curves of the newly synthetized apoproteins A-I and A-II (expressed in dpm/µmole of leucine)(Fig.3) coincide which is in favor of their secretion as a combined particle into the circulation. The maximum of the activity was not reached during the first day. A delay of 25-30 minutes is necessary before any significant activity may be recovered. This delay is the time needed for the incorporation of the labeled amino acid into the protein and the release of this protein as a lipoprotein

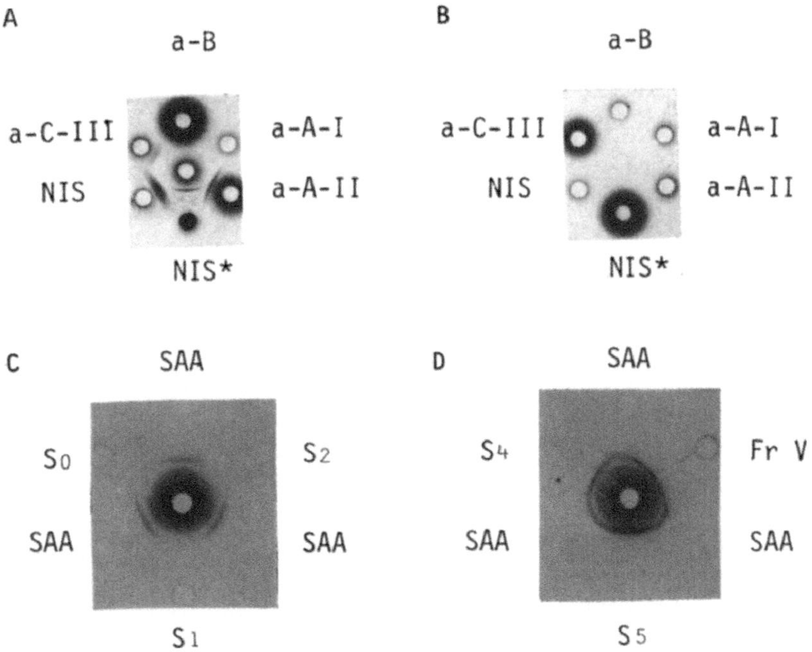

Figure 2. Double immunodiffusion analysis in 1% agar gels.

A. Centre well contains normal human plasma

B. Centre well contains apoprotein S_5

C and D) Centre well contains anti-SAA (serum amyloid A protein)

Fr V = Sephacryl Fraction V of apo HDL from a postsurgery patient

NIS = non immune rabbit serum

NIS*= the same serum concentrated 7 times

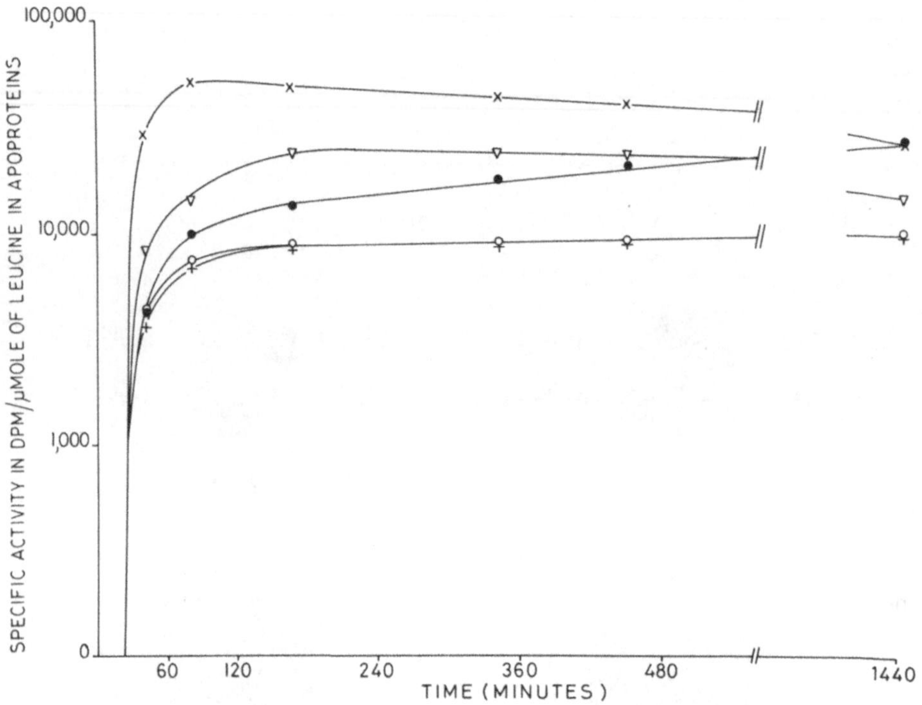

Figure 3. Specific activities (expressed in dpm/µmole of leucine) in apoproteins A-I (o), A-II (+), S_2 (V), S_4(x), S_5(o), as a function of time after (^{14}C) leucine injection. The following numbers of moles of leucine per 100,000 g of protein were chosen for apo A-I : 140, A-II :90, and S_2, S_4 and S_5 : 43, 33 and 47 respectively.

into the circulation (synthesis and secretion processes). To identify the site of synthesis, the extent of this delay is very important. As was demonstrated earlier (15) there is a 4 times greater delay for the appearance of labeled proteins when the precursor is ingested instead of intravenously injected. A similar reasoning may be applied if the label had to go to a tissue, be integrated into a membrane-binding protein and later released as fragments of this protein in the plasma upon antigenic stimulation (16). Therefore a short delay is in favor of an hepatic origin. In this patient receiving no oral feeding since 5 days and in which no exogenous source of A-I (from chylomicron) may occur, the liver is thus involved in the production of major HDL apoproteins. As there is no major difference in the extent to which isotope from labeled leucine was incorporated into apo A-I, A-II and C-II and C-III (not figured) compared to apo S, there is a presumption that all apoproteins derive from a common source, presumably the liver, without excluding however an additional intestinal production. In the animal both the liver and the intestine are known to be involved in the HDL production.

Visual analysis of the shapes of S_4 and S_5 activity curves suggest that S_5 may derive from S_4 (precursor-product relationship). More experimental evidence on a longer period of time is needed however to ascertain this hypothesis.

DISCUSSION

Three major questions arose from our observations:

1. What is the precise role of glucose infusions on the abnormally high appearance of apoprotein S in plasma?

The role of glucose in the production of abnormally high levels of apoprotein S is obvious but it may be expected that other factors are playing a role. Among these factors, the clinical status of the patient may be important although no marked and quick loss of weight may be reported. The total dependence of the patients on parenteral nutrition (intravenous perfusions) may lead to other sources of amino acids for the production of apoproteins, essentially the liver. Obese subjects put on a starvation diet for 15 days may lose up to 9% of their initial weight without showing any increase in apoprotein S levels.

The relative importance of different parameters (hypocholesterolemia, hypoproteinemia, inflammatory factors...) in addition to glucose supply has been discussed in detail in a separate paper (17).

2. Were similar proteins already reported in the literature and what may apoprotein S be compared with?

Shore et al. (18) described two threonine-poor proteins isolated from HDL in clinical conditions characterized by low levels of plasma HDL. One of these has amino acid composition, pI and molecular weight identical to apoprotein S_4. But composition, molecular weight and also partial amino acid sequence are similar to those of SAA protein (serum amyloid A) (19,20).

Besides amazing analogies, there are major differences between apoprotein S and SAA protein:

a) quantitatively, the glucose induction leads to much higher levels than any other cause of SAA protein production. Contrasting with these particularly elevated values, there is no indication in all patients studied of any obvious manifestation of primary or secondary amyloidosis;
b) the process is completely reversible and the apoprotein S returns to normal levels when the patient is discharged from the

hospital with normal laboratory tests and X rays, and a fairly good clinical condition;

c) in normal conditions apoprotein S represents only traces of HDL proteins. A normally low concentration of SAA protein has been attributed by Benditt et al. (20) to a very rapid clearance. On the contrary the specific activity curves of the more rapidly decreasing components suggest a half-life of several days, comparable to that of apo A-I and apo-II. The slow disappearance of apoprotein S from plasma observed after the change in diet in many patients, confirms this assumption.

d) the origin of amyloid A is still controversial. As discussed already in the results section, no stimulation of the immune system releasing fragments of a membrane protein in the plasma may be expected in our patients and apoprotein S is more likely a precursor than a derived fragment of a larger protein (16).

3. Do peptides S represent polymorphic forms of a true apoprotein?

The criterias required for a protein to become member of the lipoprotein family are not yet completely delineated.

"An apolipoprotein is a homogenous protein composed of a single polypeptide chain which associates with or forms an integral part of a lipoprotein and/or lipoprotein particle in plasma" (13). As one of the prominent functions of apolipoproteins is to transport lipids in a water-soluble form, it appeared essential to demonstrate at least in vitro that the protein is able to bind lipids and to form a stable complex. Segrest et al. (21) have suggested that the human plasma apoproteins contain specific phospholipid binding sites. There is an amphipathic helical region typical for each apoprotein. All apoproteins of the A and C families are known to bind phospholipid (DMPC) and phospholipidcholesterol.

Segrest et al. (16) have shown that amyloid A contains an amphipathic helical segment of 26 residues which represent an important structural unit for lipid-protein association and they, in fact, demonstrated spontaneous association of amyloid A with phospholipid vesicles and formation of a complex. Partial sequence of apo S included the same helical amphipathic segment of 26 residues.

Apo S has been already studied by physical methods, and shown to spontaneously associate with phospholipids (DMPC). At about the gel-crystalline phase transition temperature of DMPC (24°C) the association of apo S with the lipid was the most rapid (24.3^{o}C). The association kinetics is faster with apo S than with apo A-I and close to that of apo A-II. The effect of temperature and of the cholesterol content of the cholesterol-DMPC vesicles is less pronounced on the association kinetics of apo S than of apo A-I. The elution profile of the DMPC-apo S mixture on a sepharose 6 B column indicates that the isolated complex consists of 50 mol phosphatidylcholine per mole of apoprotein S. These results will be developed in a companion paper (8).

CONCLUSION

Large amounts of a new family of proteins (S) were isolated from HDL of intravenously injected neurological and postsurgery patients where they replace part of apoproteins A-I and A-II. The synthetic and catabolic rates of these polypeptides are in the range of those of classical apoproteins. The peptides bind spontaneously in vitro with phospholipid vesicles and seem to bind, in vivo as well, lipids. More information is needed before including these peptides and SAA protein into a new class of apoproteins very sensitive to both dietary and pathological alterations in humans.

REFERENCES

1. Malmendier, C.L., Christophe, J. and Amerychx, J.P. Separation and partial characterization of new apoproteins from human plasma high density lipoproteins. Clin. Chim. Acta 99:167-176, 1979
2. Malmendier, L., Paroutaud, P. and Amerychx, J.P. Partial amino acid sequence of three new apolipoproteins isolated from high density lipoproteins. FEBS lett. 109: 43-44, 1980
3. Malmendier, C.L. and Amerychx, J.P. S-peptides. New apoproteins of human high density lipoproteins induced by glucose infusions. Short communication. Clin. Chim. Acta, 111:267-270, 1981.
4. Schacterle, G.R. and Pollack, R.L. A simplified method for the quantitative assay of small amounts of protein in biologic material. Anal. Biochem. 51: 654-655, 1973
5. Manual of Laboratory Operations, Lipid Research Clinics Program, Vol. 1 " Lipid and lipoprotein analysis " DHEW Publ. Nº (NIH) 75-628, 1974
6. Schonfeld, G. and Pfleger, B. The structure of human high density lipoprotein and the levels of apolipoprotein A-I in plasma as determined by radioimunoassay. J. Clin. Invest. 54:236-246, 1974
7. Schonfeld, G.,Lees, R. S., George, P.K. and Prieger, B. Assay of total plasma apolipoprotein B concentration in human subjects. J. Clin. Invest. 53: 1458-1467, 1974
8. Rosseneu, M., van Tornout, P., Lievens, M-J., Amerychx,J.P. and Malmendier, C.L. Lipid binding properties of apo S_5 protein. Kinetics of association and complex isolation. Eur. J. Biochem. 1981. Submitted.
9. Anker, H.S. A solubilizable acrylamide gel for electrophoresis. FEBS Lett: 293, 1970
10. Albers, J.J., Warnick, G.R. and Chenng, M.C. Quantitation of high density lipoproteins. Lipids 13:926-

-923,1978
11. Schonfeld, G., Bailey, A. and Steelman, R.
Plasma apolipoprotein A-i and A-II levels in hyperlipidemi
Lipids 13 : 951-959, 1978
12. Scanu, A.M.
In: The biochemistry of atherosclerosis (Scanu, A.M. ed.), Dekker, New York, 1979
13. Alaupovic, P., Curry, M., McConathy, W. and Fesmire, J.
Electroimmunoassay of apolipoproteins occurring in high density lipoproteins of human plasma. Report of the high density lipoprotein methodology worshop. US Department of Health, NIH Publication Nº 79-1661, 1979
14. Blum, c:b., Levy, R.I., Eisenberg, S., Hall, M., Goebel, R.H. and Berman, M.
High density lipoprotein metabolism in man. J. Clin.Invest. 60:795-807, 1977
15. Malmendier, C.L., Amerychx, J.P. and van den Bergen, C.J.
Sequence of 1-^{14} C-leucine and 9,10-3H-palmitate incorporation into lipoprotein apoproteins and triglycerides in man. Proc. of 25 th colloquium on protides of the biological fluids (H. Peeters, ed.). Pergamon Press Oxford and New York, 1978, pp. 167-170
16. Segrest, J.P., Pownall, H.J., Jackson, R.L., Glenner, G.G. and Pollack, P.S. Amyloid A:
Amphipathic helixes and lipid binding, Biochemistry 15: 3187-3191, 1976
17. Malmendier, C.L. and Ameryckx, J.P.
L'apoprotéine S. Sa structure, son origine, sa fonction, sa destinée, ses implications possibles en pathologie. Exposés annuels de biochimie médicale. Paris, 1981. InPress.
18. Shore, V.G., Shore, B. and Lewis, S.B.
Isolation and characterization of two threonine-poor apolipoproteins of human plasma high density lipoproteins. Biochemistry 17: 2174-2179, 1978
19. Levin, M., Franklin, E.C., Frangione, B. and Pras, M.
The amino acid sequence of a malor non immunoglubin component of some amyloid fibrils. J. Clin. Invest. 51:2773--2776, 1972
20. Benditt, E.P., Eriksen, N. and Hoffman, J.S.
Origin of protein AA. Symposium on amyloidosis. Portugal, 23-28 September 1979
21. Segrest, J.P., Jackson, R.L., Morrisett, J.D. and Gotto, A.M.
A molecular theory of lipid-protein interactions in the plasmalipoproteins. FEBS Lett. 38: 247-253, 1974

ACKNOWLEDGMENTS

We wish to thank Dr. M.Y. Rosseneu of the Algemeen Ziekenhuis St-Jan, Brugge for her participation in the physicochemical study of the apoprotein S-phospholipid complexes which will be published in detail elsewhere.

We are grateful to Professor Earl P. Benditt and Dr. N. Eriksen of the University of Washington, Department of Pathology, for providing us with antisera to amyloid A and SAA protein preparations.

This study was supported in part by the "Fonds de la Recherche Scientifique Médicale" of Belgium, contract 3.9002.78.

ISOLATION OF TWO NONIDENTICAL POLYPEPTIDES FROM APOLIPOPROTEIN B OF HUMAN PLASMA

S.-O. Olofsson, K. Boström, O. Wiklund, U. Svanberg, and G. Bondjers

The arterial biology group, Dept. of Medical Biochemistry and Medicine I, University of Goteborg, Goteborg, Sweden

INTRODUCTION

Reports on the subunit structure and molecular weight of the major apolipoprotein of the low density lipoproteins (LDL), apoB, are contradictory (1-5). Results from immunochemical studies have suggested that apoB may contain more than one polypeptide (4). In this paper we report on the isolation and partial characterization of two polypeptides belonging to apoB.

MATERIAL AND METHODS

Solubilized apoLDL was prepared by the method described earlier (6)

Cromatographic procedures

Chromatography on Sephadex G100, DEAE-cellulose and Sepharose 6BCL was carried out as described earlier (6).

Chromatography on Sepharose 4 BCL (Pharmacia Fine Chemical, Uppsala, Sweden) was performed on a 90x2.5 cm column equilibrated with 0.01 M tris buffer pH 8.0 with 8 M urea and 5% β-Mercaptoethanol. Fractions of 4 ml were collected and analysed with polycrylamide gel electrophoresis (2.1 % acrylamide monomer see below).

Electrophoresis was carried out on polyacrylamide gels containing urea (2.1% and 7% acrylamide concentration) (7,8) or sodium dodecyl sulphate ($NaDodS_4$); (5% acrylamide concentration (9).

ISOELECTRIC FOCUSING

Flat bed isoelectric focusing was carried out on 1% IEF-agarose (Pharmacia Fine Chemicals, Uppsala, Sweden) containing 4 M urea, 12% sorbitol and ampholyte (dilution 1:16) (Pharmalyte, Pharmacia Fine Chemicals, Uppsala, Sweden). A constant power of 5w was used

and the isoelectric focusing was carried out for 3 hours + 4º.
The pH gradient was measured with a surface pH electrode (LKB, Bromma, Sweden)

Amino acid analysis was carried out by the methods used earlier (10).

IMMUNOCHEMICAL METHODS

Apolipoproteins and antiserum to apolipoproteins were prepared by the methods used earlier (6.11).

Solubilized apoB was obtained by delipidization and solubilization (6) of a narrow density cut (d 1.030-1050 g/ml) of LDL-2.

Immuno-double diffusion was performed in 1% agar gels (lipoproteins) or 1% IEF-agarose gels (purified polypeptides).

RESULTS

Isolation of apoBF-I (a high molecular weight polypeptide from apoLDL)

Apo LDL, was chromatographed on Sephadex G100 (6). The major fraction was eluted just after the void volume of the column. This fraction was chromatographed on DEAE-cellulose column. An unretained fraction was obtained. It was chromatographed on Sepharose 6 BCL. Two fractions were obtained from the Sepharose 6 BCL. The major fraction eluted in the high molecular weight region of the column contained apoBF-I. This fraction was further purified by chromatography on Sepharose 4BCL under reduced conditions.

Isolation of apoBF-II was performed with methods described earlier (cf. band I ref. 6).

Electrophoresis of apoBF-I and apoBF-II

After reduction apoBF-I migrated as one protein staining band, on electrophoresis on urea containing polycrylamide gels with 2,1% acrylamide concentration at basic PH. ApoBF-I did not migrate into separating gels with 7% acrylamide concentration.

ApoBF-II appeared as one protein staining band on electrophoresis on 7% polyacrylamide gels containing urea at basic and acid pH (6).

After reduction apoBF-I and apoBF-II appeared as single protein staining bands on electrophoresis on $NaDodSO_4$, containing polyacrylamide gels with 5% acrylamide concentration. The migration of apoBF-I correponded to an appearent molecular weight of 250 000, while apoBF-II migrated in a position correponded to a molecular weight of 70 000.

Isoelectric focusing

ApoBF-I and apoBF-II appeared as single protein staining bands on iso-electric focusing with estimated iso-electric points of 6.1 and 6.3 respectively amino acid composition.

All commonly found amino acids were present in apoBF-1 and apoBF-II.

Immunochemical studies

Solubilized apoB gave two precipitin arcs on immuno-double diffusion with antiserum to solubilized apoB. One of these corresponded to apoBF-I the other to apoBF-II.

There was total identity between the precipitin arcs formed between whole serum or LDL and antiserum to apoBF-I, Lp-B and apoBF-II, indicating that the two polypeptides accur on the same lipoprotein partocles in serum i.e. is a part of Lp-B.

DISCUSSION

The data presented in this paper indicate that apoB is heterogeneous composed of two non-identical polypeptides. One of these polypeptides.has a high molecular weight (aproximately 250 000). The presence of a polypeptide within apoLDL with a molecular weight of 250 000 has earlier been suggested (1).

The other polypeptide subunit of apoB has an apperent molecular weight of 70 000 (6). The presence of a polypeptide within apoLDL with a molecular weight of 70 000 is suggested from electrophoresis of solubilized apoLDL on polyacrylamide gels containing $NaDodSO_4$, (2-5). In non of these reports (2-5), however, have the polypeptide corresponding to the protein staining band migrating in a position correponding to 70 000 been isolated.

Based on the data presented in this paper we suggested that apoBF-I should be referred to as the B-I polypeptide and apoBF-II as the B-II polypeptide.

Acknowledgments

We thank Anita Magnusson and Margit for skillful technical assistents and Agneta Ladström for typing the manuscript.

This study was supported by grants from the Swedish Medical Research Council Nº 4531, the Swedish Oleo-Margarine Foundation for Nutritional Research, the Swedish National Association against Heart and Lung Disease and the University of Goteborg.

REFERENCES

1. Smith, R., Dawson, J.R. & Tanford, C. (1972) J. Biol. Chem. 247: 3376.
2. Chen, C.-H.& Aladjem, F. (1974) Biochem. Biophyd. Res. Commun 60:549.
3. Chen, C.-H & Aladjem, F. (1978) Biochem. Med. 19: 178
4. Huang, S.S. & Lee, D.M. (1979) Biochem. Biophy. Acta 577: : 424.

5. Socorro, L. & Camejo, G. (1979) J. Lipid Res. 20: 631.
6. Olofsson, S.-O., Bostrom, K., Svanberg, U. & Bondjers, G. (1980) Biochemistry 19:1059.
7. Davis, B.J. (1960) Ann. N.Y. Acad. Sci 121:404.
8. Reisfeldt, R.A., Lewis, U.J. & Williams, D.E. (1962) Nature (London) 195: 281.
9. Weber, K. & Osborn, M. (1969) J. Biol. Chem 244: 4406.
10. Olofsson, S.-O., Fager, G. & Gustafson, A. (1977) Scand. J. Clin. Lab. Invest. 37: 749.
11. Olofsson, S.-O., McConathy, W.J. & Alaupovic, P. (1978) Biochemistry 17: 1032.

A COMPARATIVE STUDY OF LIPIDS AND HDL-CHOLESTEROL IN OLD JOGGERS AND PATIENTS WITH CORONARY INFARCTION

W. Schwartzkopff*, K. Peslin*, F. Nussel*, C. Luley*,
W. Doehrn**, and W. Dransfeld***

Lipidmetabolic Outpatient-Department at the Klinikum
Charlottenburg - Free University of Berlin-West -
** Dusseldorf
*** BFA-Klinik Wannsee Berlin-West

INTRODUCTION

The close correlation between LDL-cholesterol concentration and risk of myocardial infarction is well known (4). LDL-cholesterol is considered an atherogenic factor (8). In contrast it would appear that HDL has antiatherogenic properties, since there is a negative correlation between HDL-cholesterol and risk of myocardial infarction (1, 2). Low HDL-cholesterol concentrations are found with high frequency in patients with infarction (3,4,15), while high HDL concentrations are associated with longevity and reduced risk of coronary artery disease (2).

HDL-cholesterol concentrations below 35 mg/dl in males and 45 mg/dl in females are considered as independant risk factors (1, 2). Therefore, prevention of cardiovascular disease should include both a reduction in serum lipids, especially in LDL--cholesterol, and an increase in HDL-cholesterol. This can be achieved through medication with Bezafibrate or Fenofibrate, Cholestyramine (7), low nicotine consumption, small amounts of ethyl alcohol and through physical activity (5, 6, 9, 10, 11, 12, 16, 17, 18, 19).

We shall deal with the influence of a jogging regimen of serum lipids in a group of men older than 60 years, and we shall compare these data with those from a group of 335 patients with myocardial infarction. We should like to put the following questions:

1. What is the influence of a jogging regimen of 10 km, 3 times per week, on serum lipids, especially LDL - and HDL-choles-

terol in a group of men aged 62 years and older?

2. What changes in concentration of serum lipids, insulin, cortisol, growth hormone (GH), glucose and free fatty acids are found immediately after a 10-Km-run?

3. Is there a difference in lipid-concentration-especially in HDL-cholesterol concentrations - between the group of joggers and a group of patients who have successfully completed a rehabilitation program 1 year after myocardial infarction?

METHODS

The data presented here were obtained from 33 male joggers ranging in age from 62 to 82 years and from 268 men and 67 women between the ages of 40 and 60 years who had suffered myocardial infarction 1 year before the start of the study. After precipitation of VLDL and LDL with Phosphotungstate acid (Boehringer-Test) the HDL-cholesterol concentration was determined in the supernate and LDL-cholesterol concentration calculated using the formula of Friedewald. HDL_2 and HDL_3 - cholesterol were also determined by preparative ultracentrifugation with separation of the sera at density gradients of 1.073 g/ml (= HDL_2) and 1.125 g/ml (=HDL_3).

RESULTS

Triglyceride values in the joggers were significantly lower than in a group of healthy controls, males and females ranging in age from 30 to 60 years (Tab. 1). There were no differences

Table 1. Lipid concentrations in healthy women and men as well as in old joggers

	TG mg/dl	total cholesterol mg/dl	HDL-CH mg/dl	LDL-CH mg/dl
WOMEN 30-60 years	75 ±19	216 ±31	62 ±9	142 ±36
MEN 30-60 years	100 ±34	210 ±28	49 ±12	144 ±26
JOGGER 61-82 years	69 ±23	226 ±27	60 ±13	152 ±23

in cholesterol values among the three groups. This was also true of LDL-cholesterol, with a mean value of 144 mg/dl in healthy males and females as well as in the group of joggers 152 mg/dl. HDL-cholesterol values were higher in the healthy females than in the group of male controls and were virtually identical with the HDL-cholesterol values found in the group of joggers.

The second question dealt with the immediate influence of a 10-Km-run on lipid and lipoprotein concentration. 24 men participated, and the fastest covered 10 Km in 43 minutes, the slowest in 60 minutes (Tab. 2). There was a significant increase in triglyceride concentration during exercise, from 77 to 105 mg/dl. Total cholesterol remained constant, while there was a significant increase in HDL-cholesterol. A low LDL-CH : HDL-CH ratio of 2.55 declined still further to 2.1 after exercise.

It was interesting to note the changes in Apoprotein-HDL, Apoprotein B and HDL_2- and HDL_3- cholesterol that accompanied these slight but unequivocal shifts in LDL - and HDL-cholesterol (Tab. 3). There was a significant decrease of 17 mg in the Apoprotein HDL during the run, while Apoprotein B concentration remained unchanged. HDL_2 - cholesterol increased from 41 mg/dl to 48 mg/dl. The increase in HDL_3 - CH - concentration was slight but not statistically significant and the HDL_3-cholesterol value of 22 mg/dl agrees with the value of 21 mg/dl recently reported by Schafer et al. We found a HDL_2 - CH -concentration of 41 mg/dl, while Schafer reported 36 mg/dl for normal untrained men.

Table 2. The lipids before and after running 10 Km between 43 and 60 min in old joggers ($* = p < 0,01$)

	$\bar{x}$	min	max
Age (years)	70.7	65	82
Weight (Kg)	65.5	55	79
Height (cm)	170.5	164	176
Ideal weight (%)	100%		

Lipids	before $\bar{x}$	min	max	after $\bar{x}$	min	max
TG (mg/dl)	77	35	145	105*	60	138
CH (mg/dl)	242	162	304	244	176	340
LDL-CH (mg/dl)	163	105	231	157*	95	257
HDL-CH (mg/dl)	62	36	91	73*	52	99

Table 3. Apoprotein-HDL and Apoprotein B, HDL_2, HDL_3-CH in old joggers before and after running 10 km's

				after		
	before	min	max	$\bar{x}$	min	max
Apo-HDL (mg/dl)	262	194	341	245*	196	311
Apo B (mg/dl)	97	57	133	102	70	131
HDL_2-CH d= 1,073 g/ml	41	26	73	48**	28	7
HDL_3-CH d= 1,125 g/ml	22	11	31	26	10	36
HDL_2/HDL_3		2,68			2,86	

* = $p < 0{,}001$, ** $p < 0{,}01$

Joggers demonstrate higher total HDL-CH concentrations as well as a shift in the relation between HDL_2-and-HDL_3-cholesterol with an increase in the former. There are reports in the literature of HDL_3-CH concentrations which exceed HDL_2-CH concentrations, and the shift from the heavier HDL_3 to the lighter and lipid rich HDL_2 is ascribed to the lipolytic properties of HDL_3(13).

HDL_2 develops from HDL_3 after meals and under a diet rich in lipids. The high values for free fatty acids found in the joggers- -800/uval/ml - were 300/uval/ml higher than those found in untrained subjects and demonstrate the high degree of lipolytic activity in the physically active subjects.

Immediately after the 10-Km-run, we measured increase in serum glucose, free fatty acids, growth hormone (GH) and cortisol (Tab.4). Lipolysis may have beeen activated by cortisol and GH to some extent, since insulin values remained unchaged.

The third point of my address concerns lipids and lipoproteins, especially HDL-cholesterol in patients who have suffered myocardial infarction (fig.1). We found normal lipid values in 45% of patients of both sexes in this group. 18% demonstrated a hyperlipoproteinemia type IIa, 15% type IIa and 20% type IV. HDL-concentrations were determined separately for males and females in the group with myocardial infarction and demonstrated a wide distribution with generally lower values than those found in the group of joggers (fig. 2). 45 mg/dl was set as lower limit of normal in females, and 35 mg/dl in males.

We investigated the incidence of normal and low HDL-cholesterol

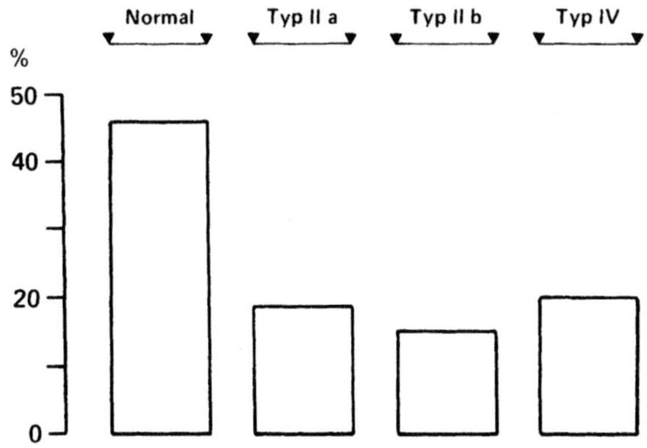

Fig. 1. Procentual distribution of the normo-and hyperlipoproteinemia type II a, II b, IV in patients with cardiac infarction (n=335)
(Schwartzkopff, Peslin, Dransfeld, 1980)

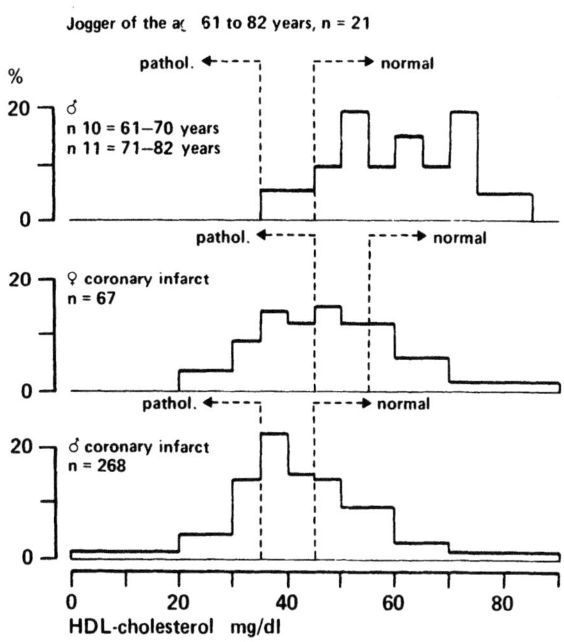

Fig. 2. Procentual distribution of the HDL-CH in dependance to its concentration

Table 4. The influence of exercise on some metabolic and hormonal parameters in old joggers

Glucose (mg/dl)	97 ± 18	133 ± 19 *
Insulin (μE/ml)	6.5 ± 9.2	6.5 ± 5.4 n.s.
Free fatty acids (μVal/l)	799 ± 266	1431 ± 740 *
Growth hormone (ng/ml)	4.2 ± 3.6	7.1 ± 7.0 *
Cortisol (μg/ml)	19.3 ± 6.2	32.4 ± 7.7 *

= $p < 0.0001$

concentrations in the patients with myocardial infarction.As the tab. 5 shows, HDL-cholesterol concentrations were higher than 41 mg/dl in 56,5% of the patients in this group. 22,9% of all patients demonstrated concentrations lower than 35 mg/dl. Only 7,5% of patients with myocardial infarction and normal serum lipid values and 8% of patients with HLP-type IIa had low HDL-cholesterol values but in 38% of patients with type IIb and in 38% of patients with type IIb and in 59% with HLP type IV there were HDL-cholesterol values lower than 35 mg/dl (Fig. 3). This investigations shows that low HDL-cholesterol values are rarely found with a normal lipid status also in patients with history of myocardial infarction. There is a negative correlation between hypertriglyceridemia and HDL-cholesterol. It is as yet unclear whether men and women with low HDL-cholesterol values are at a higher risk of myocardial infarction.

DISCUSSION

Our data show that physical training has a positive effect on serum lipid and lipoprotein values, even in older subjects. We found a clear decrease in triglyceride concentrations coupled with normal LDL and total cholesterol values and significant increase in HDL-cholesterol with a shift from HDL_3-to HDL_2-cholesterol. This demonstrates increased lipolytic activity.In patients with myocardial infarction an appropriate exercise program should not only increase coronary reserve but also help to normalize serum lipid concentrations, especially by reducing LDL-cholesterol concentration.

IN SUMMARY

1. Physical exercise increase HDL-cholesterol concentrations; the HDL_2subfraction increase significantly, while HDL_3 concentrations remains unchanged.

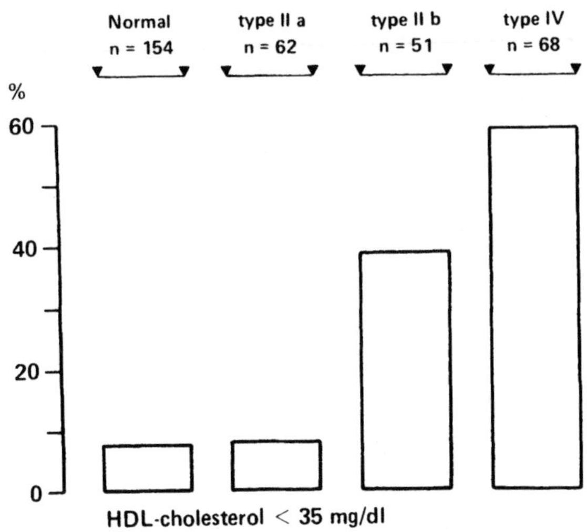

Fig. 3. Procentual incidence of a pathological HDL-CH in patients with cardiac infarction in dependence of the HLP-type

2. Exercise decrease serum triglyceride concentrations and increases lipolytic activity as a result of the increased HDL concentration.

3. 50% of all patients demonstrated hyperlipoproteinemia 1 year after myocardial infarction, with 18,5% type IIa, 15% type IIb and 20% type IV.

Table 5. HDL-cholesterol in 335 patients with cardiac infarction one year after infarction

HDL-CH mg/dl	n 335	% = 100	HLP-Typ IIa n = 62	HLP-Typ IIb n = 51	HLP-Typ IV n = 68	no HLP n = 154
>41	189	= 56,5	50	27	16	96
40-36	69	= 20,5	7	4	12	46
35-31	43	= 12,8	5	11	19	8
< 30	34	= 10,1	0	9	21	4

4. HDL-cholesterol concentration was lower than 35 mg/dl in only 7% of patients with normal serum lipoprotein values and in type IIa. Low HDL-cholesterol concentrations were found in 38% of patients with type IIb and in 59% with HLP-type IV. HDL-cholesterol concentration correlates negatively with these two groups, especially with the triglyceride concentration.

5. Further studies will show whether low HDL-cholesterol values constitute an independent risk factor for coronary artery disease.

REFERENCES

1. Assmann, G., Oberwittler, H. Schulte, H. Schriewer, H. Funke. P. H. Epping und W.H. Hauss
" Pradiktion und Fruherkennung der koronaren Herzkankheit "
Internist 446-459, 1980

2. Assmann, G., Schriewer, H. Schulte und W. Oberwittler
" Der Stellenwert des HDL-Cholesterin als Risikoindicator der Koronaren Gefasskrankheit "
Internist 21 202-212, 1980

3. Avogaro, P., G. Bittolo Bon, G. Cazzolato, G.B. Quinci
" Are Apolipoproteins better Discriminators than Lipids for Atherosclerosis "
The Lancet, Apr. 28, 901-902, 1979

4. Avogaro, P., G. Cazzolato, G. Bittolo Bon and F. Belussi
" Levels and Chemical Composition of HDL_2, HDL_3 and other Major Lipoprotein Classes in Survivors of Myocardial Infarction"
Artery 5 (6) 495-508, 1979

5. Dufaux, B., H. Liese, R. Rost, H. Heck, W. Hollmann
" Uber den Einfluss eines Ausdauertrainings auf die Serum--Lipoproteine unter besonderer Berucksichtigung der Alpha-Lipoproteine (HDL) bei jungen and alteren Personen
Deutsche Zeitschrift fur Sportmedizin, V, 123-127, 1979

6. Enger, Sven Chr., Kirsten Herbjornsen, Jan Erikssen und Ane Fretland
" High density lipoproteins (HDL) and physical activity: the influence of physical exercise, age and smoking on HDL-cholesterol and the HDL-/total cholesterol ratio "
candinavian Journal of Clinical and Laboratory Investigation 37, 251-255, 1977

7. Noseda, G., B. Lewis, R. Paoletti
" Diet and Drugs in Atherosclerosis "
European Atherosclerosis Group Meeting, 83-91, 1979

8. Goldstein, J.L., M.S. Brown
" The low-density Lipoprotein pathway and its relation to atherosclerosis"

Annual Review of Biochimie 46:897-930, 1977

9. Huttunen, J.K., E. Lansimies, E. Voutilainen, C. Ehnholm E. Hietanen, I. Pentila, O. Siltonen and R. Rauramaa " Effect of Moderate Physical Exercise on Serum Lipoproteins " Circulation 60, no. 6, 1220-1229

10. Lehtonen, A. and J. Viikari " Serum Triglycerides and Cholesterol and Serum High-DEnsity Lipoprotein Cholesterol in Highly Physically Active Men " Acta Medica Scandinavica, 111-114, 1978

11. Lopez-S., A., R. Vial, L. Balart and G. Arroyave " Effect of Exercises and Physical Fitness on Serum Lipids and Lipoproteins " Atherosclerosis 20, 1-9, 1974

12. Martin, R.B., W.L. Haskell and P.D. Wood " Blood Chemistry and Lipid Profiles of Elite Distance Runners " Annals New York Academy of Sciences, 346-360

13. Patsch, J.R., A.M. Gotto Jr. " Die Rolle von " High-Density "- Lipoproteinen (HDL) im Katabolismus triglyceridreicher Lipoproteine " Biochemie, Klinik, Epidemiologie, 17-22, 1979

14. Schafer, E.J., D.M. Foster, L.L. Jenkins, F.T. Lindgrem, M. Berman, R.I. Levy and H.B. Brewer, Jr. " The Composition and Metabolism of High Density Lipoprotein Subfractions " Lipids, Vol. 14, Nr. 5, 511-522

15. Vergani, C., Trovato and Nicola Dioguardi " Serum total Lipids, Lipoproteins Cholesterol, Apoproteins A and B in Cardiovascular Disease " Clinica Chimica Acta, 87, 127-133, 1978

16. Wood, P.D. and W.L. Haskell " The Effect of Exercise on Plasma High Density Lipoproteins " Lipids, Vol 14, No. 4, 417-427

17. Wood, P.D., W.L. Haskell, M.P. Stern, S. Lewis and C. Perry " Plasma Lipoprotein Distributions in Male and Female Runners " Annals New York Academy of Science, 749-763

18. Wood, P.D., W. Haskell, H. Klein, S. Lewis, M.P. Stern and John W. Farquhar " The Distribution of Plasma-Lipoproteins in Middle-aged Male Runners " Metabolism, Vol 25, No. 11, 1249-1257, 1976

19. Dufaux, B., Assmann, G., Hollmann, W.
" Biochemische Aspekte der Lipoproteinveranderungen bei Sportlern " (Biochemical Aspects of Lipoprotein Changes in Athletes)
Muncher Medizinische Wochenschrift, 112 (1980) Suppl. 5, 244-250

PHOSPHOLIPIDS AND PLATELET AGGREGATION

L. Douste-Blazy, G. Mauco, B. Perret, M. Plantavid,
F. Laffont, M.F. Simon, and H. Chap

Inserm Unité 101, Biochimie des Lipides, Hôpital Purpan
31059 TOULOUSE, France

INTRODUCTION

Platelet is the subject of more and more numerous studies, justified in the first place by its essential role in haemostasis and coagulation, and consequently in the amplification of the responses characterizing thrombosis. The phospholipids are tightly linked to these phenomena.First, they play a mainly structural role because of their particular physicochemical properties which allow the formation of the phospholipid double-layer typical of the biological membranes. They also have a more dynamical contribution to platelet physiology, through the activation of their metabolism during the platelet physiology, through the activation of their metabolism during the platelet stimulation, the release of arachidonic acid and the procoagulating properties of certain of them, such as phosphatidylserine. At last, a phospholipid structure was recently discovered for a potent platelet aggregating factor, the PAF-acether.

PLATELET MEMBRANE PHOSPHOLIPIDS

In our laboratory, the group of Hugues Chap (1) has specified the phospholipid organization in human platelet membranes. Using a new method of membrane preparation (2) based on the utilization of a specific membrane marker (tritiated concanavalin A) and a new separation medium (Percoll), we have performed a non-lytic degradation of platelets, coupled to a gas chromatography analysis of the fatty acids liberated by phospholipase A_2.

A phospholipid asymmetry was observed in the membrane. The most important point is the almost exclusive location of phosphatidylserines inside the platelet. Because these phospholipids have a strong procoagulating capacity (platelet factor 3), this peculiar distribution gives an explanation to the latency of the procoagulating activity of nonactivat2d platelets (3).

The arachidonic acid is not homogeneously distributed since 90% of the membrane arachidonic acid are present in the inner leaflet. This fatty acid being liberated from phospholipids during the aggregation to be converted into thromboxane A_2, this peculiar organization consequently has an important physiological significance.

PHOSPHOLIPID METABOLISM DURING AGGREGATION

Several studies take into account the modifications of the phospholipid metabolism during the platelet activation. So, a very fast increase of ^{32}P-orthophosphate incorporation into phosphatidic acids was described. This phenomenon seems to correspond to the shape change induced by ADP in the platelet, but it is much more important during stronger stimulations producting the release reaction

The metabolic accelerations are very interesting in so far as they fit in the phospholipid effect described for many tissues during their stimulation. For Michell (4), these modifications would be consecutive to the activation of phospholipase C specific for the monophosphatidylinositols, the diglycerides produced being immediately rephosphorylated by the diglyceride-kinase.

We have shown that the phospholipase C from Clostridium welchii induced an aggregation (5) and a platelet secretion and we have studied the modifications of the phospholipid metabolism induced by this enzyme. ^{32}P-ortophosphate prelabelled platelets were incubated in the presence of phospholipase C at aggregating concentrations. Under these conditions, a very important stimulation of the phosphatidic and lysophosphatidic synthesis was obtained.

The phosphatidic acid appearance corresponds to the phosphorylation of the diglycerides formed in the membrane under the effect of the platelet diglyceride-kinase. The absence, in the platelet lysates, of a phospholipase activity towards the phosphatidic acids led us to propose that a diglyceride lipase and a monoglyceride kinase could successively interfere, the latter enzyme having been described in the platelets.

Using tritiated diolein, we have shown the existence in human platelets of a diglyceride lipase which was active at optimum pH 7 (6). Under our experimental conditions, the monoglycerides did not accumulate and were rapidly deacylated. On the other hand, using the subcellular fractionation technique on Percoll, as described above, we have established the membranous localization of this enzyme. This suggests a possible regulation by the monoglyceride - Kinase of the 2-monoglyceride deacylation and consequently of the release of the arachidonic acid, a precursor of thromboxane A_2.

Moreover, we have shown that a physiological aggregating agent, thrombin, stimulates the production of phosphatidic and lysophosphatidic acids in platelets. The similitary of these modifications with those induced by phospholipase C led us to describe a phospholipase C activity in platelet cytosol.

THE PHOSPHOLIPIDS AS AN ARACHIDONIC ACID " RESERVOIR "

Only the free form of arachidonic acid is metabolized by the cyclo-oxygenase (8). Non-esterified arachidonic acid is practically non-existent in the platelets, whereas almost all this fatty acid is located in position 2 of the phospholipids. Consequently, it is generally accepted that one of the key steps of the platelet activation is the release of arachidonic acid from the phospholipids. Consequently, it is generally accepted that one of the key steps of the platelet activation is the release of arachidonic acid from the phospholipids.

The incorporation of radioactive arachidonic acid into the platelets is followed by a transfer of the radiactivity from the phospholipids to the free arachidonic acid and its derivatives, upon exposure to thrombin (9). The phospholipids involved are lecithins and monophosphatidylinositols (9) (10). For some authors however (11), the arachidonic acid-producting phospholipids would be phosphatidylethanolamines, a phospholipid particulary rich in arachidonic acid and a preferential substrate for endogenous phospholipase A_2 (12).

If the importance of arachidonic acid release is obvious, the controversy remains entire about the nature of the phospholipids, the enzyme and even the subcellular localization of their mechanism. It is generally admitted that arachidonic acid release is performed by an endogenous phosphalipase A_2 and several authors have measured such an activity without entirely identifying these enzimes (13) (14). Derksen and Cohen (11) have suggested a membrane inner localization for phospholipase A_2. Béréziat et al. (15) have shown a phospholipase A_2 activity towards 2-arachidonyl-lecithin, in the dense granules and in the plasma membrane. We have pointed out a possible alternative pathway which would involve the sequential action of two enzimes, a phospholipase C and a lipase, to release the arachidonic acid of platelet phospholipids.

ANOMALIES OF MEMBRANE PHOSPHOLIPIDS IN PLATELETS

We have observed anomalies of platelet membrane phospholipids in two cases of Bernard-Soulier disease and among patients having suffered from cerebral thrombosis.

In collaboration with the group of Prof. J. Caen (Hôpital Lariboisière, Paris), we attempted to detect some anomalies of membrane phospholipids in the platelets of patients affected by the Bernard-Soulier disease. This congenital disease responsible for haemostasis disorders shows in particular a deficiency in certain membrane glycoproteins. We first observed a modification in the phospholipids composition of these membranes, characterized by an accumulation of lysophospholipids, known for their ability to destabilize the phospholipid double layer.

On the other hand, the phospholipid distribution between the two leaflets of these membranes is modified. The asymmetric distribution that we demonstrated in normal cells does not seem to be so well preserved, since for instance the procoagulating phospholipids such as phosphatidylserines almost exclusively directed to the inside of the normal cell, seem to be more exposed towards the plasma compartment in the case of these pathological platelets. this study contributes to explain the increase of platelet factor 3 observed in Bernard-Soulier platelets. Moreover, this phospholipid anomaly sets the problem of the part played by the lipido-protidic interactions in the preservation of membrane structure and function.

For thrombosis a model of aggregation to phospholipase A_2 was perfected and constitutes a functional test to explore the arachidonic acid amounts present on the outer layer of the membrane. The comparison of this aggregation in a sample of patients suffering from cerebral thrombosis to that of normal subjects revealed that there was no aggregation to phospholipase A_2 among 50% patients and only 25% control subjects. As all of them normally aggregate to arachidonic acid, it can be thus concluded that the surface arachidonic acid diminishes in platelets from patients. The origin and biological significance of this decrease should be worth investigating.

PAF-ACETHER AND PLATELET AGGREGATION

PAF (platelet activating factor) was found in 1972 (16) as originating from circulating basophilic polynuclears in rabbit blood and responding to an antigen specific for their membrane IgE. It was proved, on several animal species, that it could also be secreted by polymorphonuclear leukocytes, monocytes and macrophages (17). It triggers the platelet aggregation of various animal species at molar concentrations ranging from 10^{-10} to 10^{-12} M (18) (19). It also acts on human platelets and it was recently demonstrated that platelets themselves can release PAF-acether (20). Its chemical structure was elucidated; it is a phospholipid: 1-o-alkyl-2-acetyl--glycero-3-phosphorylcholine (21) (22).

PAF-acether is the first example of a phospholipid mediator and can be considered as the third mediator of platelet aggregation, after ADP and thromboxane A_2. Like TxA_2 and ADP, PAF is a substance released by platelets under the action of various agents such as collagen or thrombin, in the presence of calcium ions. Like serotonin and histamine, PAF-acether is a mediator of inflammation and broncho-constriction (23). This new mediator is particularly interesting because being released by the platelets themselves, it seems to interfere in thrombosis formation.

We recently showed (24) that rabbit platelets activated by Ca-ionophore A 23187 incorporated tritiated acetate in a phospholipid which, by thin-layer chromatography, exhibited an Rf identical to that of PAF-acether. After HPLC purification, this product was submitted to various treatments (alkaline methanolysis, acidic hydrolysis, incubation in the presence of lipase and phospholipase A_2) which revealed that it was actually a mixture of 1-acyl-2-acetylglycero-3-phosphorylcholine and 1-alkyl-2-acetyl-glycero-3--phosphorylcholine. A selective hydrolysis of the first component by Rhizopus arrhizus lipase does not modify the mixture aggregating activity, which can be thus attributed almost totally to 1-alkyl phospholipids. Consequently, we suggest that the calcium intracellular increase would activate a platelet phospholipase A_2, thus generating 1-alkyl and 1-acyl lysophosphatidylcholines used as acceptors for acetyl group.

This preliminary study represents the first biochemical characterization of PAF-acether function and starts the identification of enzymes implied in this synthesis.

So, the membrane phenomena which develop in the platelets are fundamental in the interactions with endothelium, coagulation factors or at the time of platelet activation.

Platelets remain essential elements in the genesis of thrombotic diseases, one of the major pathologies as well by its severity and suddenness as the number of patients who are affected.

REFERENCES

1. Chap, H., Zwaal, R.F.A. & Van Deenen, L.L.M.,Biochim. Biophys. Acta, 1977, 467, 146-164.
2. Perret, B., Chap, H.J. & Douste-Blazy, L., Biochim. Biophys. Acta, 1979, 556, 434-446.
3. Zwaal, R.F.A., Comfurius, P. & Van Deenen, L.L.M., Nature, 1977, 268, 358-360.
4. Michell, R.H., Biochim. Biophys. Acta, 1975, 415, 81-147.
5. Chap,H. & Douste-Blazy, L., Eur. J. Biochem., 1974, 48, 351-355.
6. Mauco, G., Chap, H., Simon, M.F. & Douste-Blazy, L., Biochimie, 1978, 60, 653-661.
7. Mauco, G., Chap, H. & Douste-Blazy, L., FEBS Lett., 1979, 100, 367-370.
8. Lands, W.E.M. & Samuellsson, B., Biochim. Biophys. Acta, 1968, 164, 426-429.
9. Bills, T.K., Smith, J.B. & Sylver, M.J., Biochim, Biophys. Acta, 1976, 424, 303-314.
10. Blackell, G.I., Duncombe, W.G., Flower, R.J., Parsons, M.F. & Vane, J.R., Brit. J. Pharmacol., 1977, 59, 353-366.
11. Derksen, A. & Cohen, P., J. Biol. Chem., 1975, 250, 9342--9347.

12. Rittenhouse-Simmons, S.& Deykin, D., Biochim. Biophys. Acta, 1978, 543, 409-412.

13. Jesse, R.L. & Cohen, P., Biochem. J., 1976, 158, 283-287.

14. Pickett, W.C., Jesse, R.L. & Cohen, P., Biochim. Biophys. Acta, 1977, 486, 209-213.

15. Trugnan, G., Bēréziat, G., Manier, M.C. & Polonovski, J., Biochim. Biophys. Acta, 1979, 573, 61-72.

16. Benveniste, J., Henson, P.M. & Cochrane, G.C.J., J. Exp. Med., 1972, 136, 1356-1377.

17. Mencia-Huerta, J.M. & Benveniste, J., Eur. J. Immonol., 1979, 9, 409-415.

18. Benveniste, J., Le Couedic, J.P. & Kamoun, P., Lancet, 1975, 1, 344.

19. Fesus, L., Csaba, B. & Muszbek, L., Clin. Exp. Immunol., 1977, 27, 512-515.

20. Chignard, M., Le Couedic, J.P., Tence, M., Vargaftig, B.B. & Benveniste, J., Nature, 1979, 79, 799-800.

21. Demopoulos, C.A., Pinckard, R.N. & Hanahan, D.J., J. Biol. Chem., 1979, 254, 9355-9358.

22. Benveniste, J., Tence, M., Varenne, P., Ricault, J., Boullet, C. & Polonovski, J. C. R. Acad. Sc. Paris,1979, 289, 1037-1040.

23. Vargaftig, B.B., Lefort, J., Chignard, M., & Benveniste, J., Eur. J. Pharmacol., 1980, 65, 185.

24. Chap, H., Mauco, G., Simon, M.F., Benveniste, J. & Douste--Blazy, L., Nature, 1980,289, 312-314.

AN APPROACH TO ANTIARTERIOSCLEROTIC PHARMACOLOGY

J. Segarra Domenech and Santafe Oroz

Departamento de Farmacologia
Colegio Universitário de Alava
Vitoria Universidad del Pais Vasco-España

INTRODUCTION

The results reached in large preventive studies carried out between 1970-1975 (Physicians of the New Castle upon Thyne region, 1971; Research Committee of the Scotish Society of Physicians 1971, The Coronary Drug Project 1970-1975) placed in the foreground, the real value of the antiarteriosclerotic drugs, profusely utilized during many years, although the conclusions have been able to be discussed, it exists a generalized opinion that present available antiarteriosclerotic drugs are not the more suitable and, consequent-ty, it is necessary a restatement of the therapeutic chapter on grounds of the new knowledges.

In this revision, the different possibilities of the antiarte-riosclerotic pharmacologic actions, are analized. Out of these comments there remain aspects related to risk factors as well as those problems in relation with compounds able to increase different lipoproteic fractions like, for instance, hypoglycemic sulfonamids, birth control drugs, etc.

DESCRIPTION OF THE POSSIBLE PHARMACOLOGIC ACTIONS

On table N-1 are collected the options of the pharmacologic actions studied experimentally at clinical levels, as well as some actions, still with theoretical characteristics.

I. Inhibition of cholesterol Synthesis

The inhibition of cholesterol synthesis, in the anaerobic or aerobic phase, was the therapeutic intent initiated some time ago and then abandoned because of its poor efficacy and for the impor-tant pharmacopathology originated by the substances utilized. Re-cently it has been considered again this pharmacologic possibility

(Endo and cols, 1976, 1979; Kuroda and cols. 1977; Tsukita and cols. 1979), by means of the use of Monacolin K and Compactin, which mechanism of action is based on the HMG-CoA reductase inhibition at hepatic level, rate-limiting enzyme in the Mevalonate synthesis. Besides the hypocholesterolemic effect, both compounds are lacking of important toxicity. The clinical studies carried out do not allow to establish its therapeutic value yet.

In relation with this therapeutic possibility it must be kept in mind:

1. Cholesterol is necessary for the synthesis of different compounds and cellular membranes in the organism, and
2. the possible repercussions that may be able to originate the HMG-CoA reductase inhibition at the cellular level of the arterial wall.

II. Action on Hyperlipoproteinemia

From the therapeutic view point the systematization of hyperlipoproteinemia (Fredrickson and Lees, 1965; Fredrickson and cols, 1977) has undoubtly advantages because allows the individualization of action on triglycerides, cholesterol, or both; but this is not a determinative factor because, at the end, what is actually important is to modify favourably the plasmatic lipoproteins by means of drugs. The pharmacologic action hiperlipoproteinemia can be sistematized in the two wide chapters:

A. Blocking the cholesterol absorption, and,
B. Acting on lipoproteins at quantitave and qualitative levels.

A. Inhibition of the Cholesterol intestinal absorption.

Although in theory, this pharmacologic procedure is valid, the resins of ionic exchange (Colestiramine, Colestipol) Neomycin and Betasitosterol pratically have not shown a great usefulness. There exist few studies on the preventive value of these compounds (Dorr and cols. 1975) and the cost, pharmacopathology, as well as the need of administering large daily doses are the principal inconvenients for their use. In the United States, during the last years resins of ionic exchange associated with Nicotinic acid and Neomycin had been used again (The Medical Letter, 1980) it seems with the scope of avoiding the increase of cholesterol produced by the resins.

B. Action on plasmatic lipoproteins

The mere plasmatic cholesterol and triglycerides reduction had been the antiarteriosclerotic therapeutic rule during many years, but recently it has begun to be considered the suggestion of Barr and cols (1951) from which arteriosclerosis was related intensively with a type of plasmatic lipoproteins. Miller and Miller (1975) verified this forgotten piece of information showing the relation between serum LDL-CO and HDL-CO with arteriosclerosis. The trigger activity of the former and the protective action of the latter on the arteriosclerotic process, have updated the importance that the appropriate modification of the different serum lipoproteins can have from the therapeutic point of view.

1. VLDL reduction. There exists a number of substances that are able to reduce plasmatic triglycerides (Nicotinic Acid, Clofibrate, Halofenate, Tiadenol, Bezafibrate, Procetofen, Metformin, Glycosaminlycans) with a greater or minor intensity. However, there is not a general consensus about the transcendency of serum VLDL reduction, because occasionally those substances may simultaneously originate an increase of LDL-CO and also of HDL-CO with eventually an opposite meaning.

2. LDL-CO reduction. The existence of a direct relation between LDL-CO and the manifestation and intensity of arteriosclerosis, makes the reduction of these lipoproteins the optimum aim of the pharmacologic action. Clofibrate and other related substances reduce LDL-CO in many patients, but occasionally an increase is verified connected with the strong VLDL reduction that these substances produce, thus the net effect on serum Cholesterol is minimum. Nicotinic acid reduces LDL-CO needing high doses that produce an important variation on type II hyperlipoproteinemias Probucol reduces LDL-CO in 10-15% when is associated with an appropriate diet (Levy 1980).

3. HDL-CO increase. For the protective character presently attributed, the increase of this cholesterol fraction by means of drugs is considered now another important goal. In this sense there exist different options.

a. Increase of HDL tissue synthesis. Although Clofibrate and Nicotinic acid increment HDL-CO (Pichardo and cols.1977), they do it with low intensity as well as it occurs with Procetofen (Rossner and Oro, 1978). At experimental level Gemfibrozil (Rodney and cols. 1976), Gemcadiol and Adamantiloxiphenilpiperazine (Maxwell and cols. 1978) intensely increase HDL-CO. According with this therapeutic possibility it has to be considered the search of substances able to modify favourablement HDL_2 fraction which protective significance is well established (Avogaro and cols., 1979).

b. LCAT induction. The important role that this enzyme exerts in the centripetal cholesterol transport mediated by HDL/Glomset, 1979) makes especially interesting the possibility of increasing its synthesis. The hepatic induction of LCAT has merited a certain attention although not uniform results have been obtained. Russel and cols. (1976) succeeded in obtaining the hepatic induction in rats, but Heller (1979) did not, either in guinea pigs or in man. These last results are in contrast with two clinical facts, possibly related with LCAT enzymatic induction:

1. Nikkila and cols. (1978) shown in epileptic patients, treated with the enzymatic inducer drug Phenytoin, an increase of HDL-CO of 33% in males and 23% in females.

2. Livingstone (1976) pointed out the low incidence of myocardial infarcts among epileptic patients. The passible therapeutic orientation of the hyperlipoproteinemia requires to go deeply in the enzymatic induction phenomenon of LCAT for what it is essential to know all the details of its synthesis.

c. Lipoprotein-lipase activity. According to the function

that is attributed to LPL in the lipoprotein interconversion, the stimulus of that enzyme opens a therapeutic possibility although some aspects related to it are still unknown, i.e., its presence in some types of hyperlipoproteinemias or the role and meaning of the enzyme of hepatic origin (H-TGL).

The physical exercise increases the lipoprotein-lipase activity. it can be assumed that LDL-CO reduction and the HDL-CO increase during this activity (Wood and Haskell, 1979; Taskinen and cols., 1980) is related to LPL. On the other hand it is known that Clofibrate (Ganesen and cols., 1976; Boberg and cols., 1977) (Ehnholm and cols., 1975) and Bezafibrate (Agustin and Greten,1979) are able to enhance the lipoprotein-lipase activity, Heparin and other glycosaminglycans markedly have the capacity to stimulate the activity of above mentioned enzyme (Zilversmith, 1973).

4. Qualitative modification of lipoproteins.This possibility of pharmacologic action has merited special attention. It is necessary to distinguish two aspects:

a. Structure modification. Sirtori and cols. (1977) as well as Rodriguez and cols. (1977) have observed that Metformine-hypoglycaemic drug - is able to avoid the appearance of arteriosclerotic lesions in rats fed with a rich cholesterol diet. This drug does not produce hypocholesterolemic effects at blood level, but reduces the ApoE content of the lipoproteins LDL, VLDL and HDL, increasing the phosphatidilcoline/esphingomyeline relation.

The clinical use of Metformine (Conti and cols., 1976; Sirtori and cols., 1977; Tiengo and cols. 1978; Descovic and cols., 1978) have demonstrated that these substance have an hypotriglyceridemic effect without modifying substantially the values of cholesterol. In none of the mentioned studies the repercussion of lipoprotein changes had been evaluated.

b. Modification of the surface electric charge of the lipoproteins. Although there is not unanimity on this respect, the surface electric charge of the lipoproteins is considered of paramount transcendency with regard to the arterial wall captation. Hoff (1979) made a pathogenic hypothesis according to which the electronegative character of the lipoprotein surfaces constitutes a risk factor. That interpretation is based on: a) the existence, in the intima of normal arteries and in the atheroma plaque, of a type of LDL with a negative charge, superior to that of the same lipoproteins in plasma(Hoff and cols., 1979; Hollander and cols., 1976, 1979), and b) "in vitro" the macrophages intake more intensively the LDL with greater negative charge secondary to its acetilation (Goldstein and cols., 1979) or after treatment with malondialdehyde.

On the contrary there are experiencies in which the intake of LDL by macrophages and smooth muscle cells is more intense when the electropositivity of its surfaces is increased after desialization (Filipovic and cols., 1979). Moreover (1975) as well as Avila and cols (1978) have observed an interaction between LDL and glycosaminglycans of the human aorta, which intensity is directly related to the electropositivity of the lipoproteins. From these results it can be assumed that changes of the electric charge on the lipoproteins surface, modify their behaviour, but it remains without elucidating which of them is the most decisive. On the other hand, it does not exist a practical verification about the meaning of those alterations.

5. Blockade of LDL in blood. The role of glycosaminglycans in the arteriosclerotic process has been frequently treated since some years ago. Its localization in the interstitial space and its reticular distribution in the arterial wall, seems to suggest an important meaning due to the following facts:

a. Glycosaminglycans bind specifically to lipoproteins apoB (LDL and VLDL), (Camejo and cols., 1975) originating soluble and non-soluble complexes (Berenson and cols., 1971; Srinivasan and cols., 1972). This interaction is more relevant with LDL of human origin.

b. As a consequency of the interaction among glycosaminglycans of the arterial wall and plasmatic LDL, accumulation of the last in the intima is produced (Iverius, 1972, 1973).

c. Smith (1977) had demonstrated that the major accumulation of LDL takes place in areas of atheromatose plaque progression and that these lipoproteins are associated with glycosaminglycans.

These facts together with the glycosaminglycans capacity for impeding the link of VLDL and LDL to their cellular receptors (Day and cols., 1975; Sirtori and cols., 1976) as well as for liberating them after the linkage (Goldstein and cols., 1976) had conducted to its therapeutic utilization. When administered, glycosaminglycans create complexes with circulating lipoproteins apoB, blocking them and avoiding their deposit in the arterial wall. At the same time they determin a relative increase of the ratio HDL Cholesterol/LDL cholesterol. This last aspect relates and reinforces the increase in absolute values of the lipoprotein HDL-CO produced by the glycosaminglycans able to activate the lipoprotein--lipase.

c. Increase of cholesterol biotransformation and excretion. This therapeutic possibility implies two aspects: a) the HDL participation in the cholesterol excretion, and b) the role of the hepatic parenchyma. The first point-even with the existing lack of knowledge in this regard - is related with the above mentioned pharmacological increase of HDL. With regard to the increase of the biotransformation and excretory hepatic capacity by means of drugs, it exists only few information on this point, and it is related to the biliary excretion of cholesterol as consequency of the sequestrum of biliary acids produced by the ionic exchange resins.

At this point, it is necessary to keep in mind that it does not exist a direct relation between the enhance of liver biotransformation capacity and its excretory function.

III. Other possibility of antiarteriosclerotic pharmacologic action. There exist several options with different meanings:

A. Modification of the arterial wall metabolism. The knowledge about the active participation of the arterial wall on the development of arteriosclerosis allows to consider these aspects from therapeutic or pharmacologic view points. The pharmacology action, at this level, is focused in two concepts:

1. Nucleotides metabolism. Shimamoto (1974) and Numano and cols., (1976), among others, consider that cyclic nucleotides are involved in atherosclerosis manifestations and this led to the introduction of Ftlazinol, a phosphodiesterase inhibitory drug, with scope of incrementing the AMPc concentration in the arterial wall. According with the opinion of some authors AMPc might be diminished in cases of atherosclerosis.

2. The ratio free cholesterol/sterified cholesterol in the arterial wall. At the beginning of the century, Windaus (1910) pointed out that in arteriosclerotic lesions exist an increase of the mentioned ratio when compared with normal arteries. This is considered a constant metabolic data of the arteriosclerosis (Kritchevsky, 1979). At experimental level, Kritchevsky and cols. (1975) had demonstrated that Nicotinic acid, Betasitosterol, Clofibrate and D-tiroxine, revert the values of that ratio without modification of the plasmatic cholesterol levels. From these two pharmacological options, Ftlazinol did not give any improvement to the arteriosclerosis treatment and the Kritchevsky's theory, has not reached a pratical reality.

B. Modification of the cellular biochemical processes of the arterial wall. The pharmacological action on the participation of LDL receptors and intracellular process that follow the uptake of the mentioned lipoproteins, had scarcely been observed regarding the capacity of some glycosaminglycans to prevent LDL and VLDL bind to those receptors, as well as the capacity of liberating them from their binding (Goldstein and cols. 1976).

It is partially or totally unknown the therapeutic possibility of blocking and stimulating LDL receptors as well as the intracellular process (lysosomes, enzymes, feedback mechanisms, etc) that follow LDL captation.

c. Modification of platelet aggregation. Related with the use of antiplatelet drugs - a therapeutic approach frequently mentioned at present - it is necessary to elucidate whether the antiaggregant drugs have an antiarteriosclerotic activity or actually, they avoid the complications of the arteriosclerotic process.

c. Modification of the smooth muscle cells proliferation in the arterial wall. It has not been paid much attention, until now, to this therapeutic possibility. Epidemiologic studies, mainly published between 1970 and 1975, had demonstrated the limited efficacy of drugs used in that period. Since 1975 until nowadays, there have been introduced some new drugs of therapeutic interest, but none of them seems to be clearly better compared with the old one. These drugs have not been properly studied to evaluate their therapeutic activity. Nevertheless continuous progress in the knowledge of arteriosclerosis, specially regarding the different roles of the lipoproteins and cholesterol uptake and elimination by the peripheral tissues, allow to predict new therapeutic alternatives for which the time will established their real efficacy, but none of them seems to be clearly better compared with the old one. These drugs have not been properly studies to evaluate their therapeutic activity. Nevertheless continuous progress in the knowledge of arteriosclerosis, specially regarding the different roles of the lipoproteins and cholesterol uptake elimination by the peripheral tissues, allow to predict new therapeutic alternatives for which the time will establish their efficacy.

We acknowledge Miss Montejo for the preparation of the manuscript.

SYSTEMATIZATION OF THE POSSIBLE ANTIARTERIOSCLEROTIC PHARMACOLOGICAL ACTIONS.

I. CHOLESTEROL SYNTHESIS INHIBITION.
II. ACTION ON HYPERLIPOPROTEINAEMIAS
- A. INHIBITION OF CHOLESTEROL INTESTINAL ABSORPTION
- B. ACTION ON PLASMATIC LIPOPROTEINS
  - 1. VLDL REDUCTION
  - 2. LDL REDUCTION
  - 3. HDL INCREASE
    - a. INCREASE OF ITS SYNTHESIS
    - b. LCAT INDUCTION
    - c. LIPOPROTEIN-LIPASE ACTIVATION
  - 4. QUALITATIVE MODIFICATION OF LIPOPROTEINS
    - a. MODIFICATION OF ITS STRUCTURE
    - b. MODIFICATION OF THE ELECTRIC CHARGE OF ITS SURFACE
  - 5. BLOCKADE OF LDL IN BLOOD.
- C. INCREASE OF CHOLESTEROL BIOTRANSFORMATION AND EXCRETION

III. OTHER POSSIBLE ACTIONS.
- A. ON THE ARTERIAL WALL METABOLISM
- B. ON THE CELLULAR BIOCHEMICAL PROCESS OF THE ARTERIAL WALL
- C. ON THE PLATELET AGGREGATION
- D. ON THE PROLIFERATION OF THE SMOOTH MUSCLE CELLS OF THE ARTERIAL WALL.

REFERENCES

PHYSICIANS OF THE NEW CASTLE UPON THYNE REGION
Br. Med. J., 4: 767 (1971)

RESEARCH COMMITTEE OF THE SCOTISH SOCIETY OF PHYSICIANS.
Br. Med. J., 2: 775 (1971)

THE CORONARY DRUG PROJECT.
J.A.M.A., 214: 1303 (1970)

THE CORONARY DRUG PROJECT
J.A.M.A. 220; 996 (1972)

THE CORONARY DRUG PROJECT
J.A.M.A. 231: 360 (1975)

ENDO A., KURODA M., and TSUJITA Y.J.
Antiotics 29:1346 (1976)

ENDO A.J.,
Antibiotics 32: 852 (1979)
KURODA M., TANZAWA K., TSUJITA Y., and ENDO A.
Biochim. Biophys. Acta. 489:119 (1977)
TSUJITA Y., KURODA M., TANZAWA K., KITANO N., and ANDO A.
Atherosclerosis 32:307 (1979)
FREDRICKSON D.S., LEVY R.I. and LEES R.S.
New Engl. J. Med. 276: 32, 94, 148, 215, 273 (1976)
KRITCHEVSKY D.
Atherosclerosis. V. Proc. Five Intern. Symp., p-71. Ed.
GOTTO, Jr., A.M. Smith, L.C. and Hallen N., Springer Verlag
New York, 1980
DORR A.E., MARTIN W.B., and FREIBURGER W.A.
Adv. Exp. Med. Biol. 63:447 (1975)
THE MEDICAL LETTER
22:65 (1980)
BARR D.P., RUSS E.M., and EDER H.A.
Am. J. Med. 11: 480 (1951)
MILLER G.J., and MILLER N.E.
Lancet 1: 16 (1975)
LEVY R.I.,
The pharmacological Basis of Therapeutics. 6th. p-834.Ed.
Goodman Gilman, A., Goodman L.S. and Gilman, A. MacMillan
Publishing Co. Inc. New York. 1980.
PICHARDO R., BOULET L., and DAVIGNON J.
Atherosclerosis 26: 573 (1977)
ROSSNER S., and ORO L.
International Conference on atherosclerosis p-135. Ed.
Carlsson, L.A. Raven Press. New York. 1978.
RODNEY G. P., UHLONDERF, P., and MAXWELL.
R.E. Proc. Roy. Soc. Med. 69 suppl. 2:6 (1976)
MAXWELL R.E. NAWROCKI J.W. and UHLENDORF P.
Artery 4:303 (1978)
AVOGARO P., CAZZOLATO G., BITTOLO BON G., and BELUSSI F.
Artery 5:495 (1979)
GLOMSET J.A.,
Progress Int. Biochemical Pharmacology. vol. 15. p-41
Ed. Eisenberg, S. Karger, Bassel. 1979
RUSSEL R.L., SOLER ARGILAGA C., and HEMBERG M.
Life Sci. 19: 1247 (1976)
HELLER F.R.,
Artery 5: 110 (1979)
NIKKILA E.A., KASTE M., EHNHOLM C., and VIIKARI J.
Br. Med. J. 2:99 (1978)
LIVINGSTONE, S.
Br. Med. J., 1:586 (1976)
WOOD P.D., and HASKELL W.L.,
Lipids 14: 417 (1979)
TASKINEN M.R., NIKKILA E.A., REHUNEN S., and GORDIN A.
Artery 6: 471 (1980)

GANESA D., WHINE T.F., and VITIAK J.R.
Artery 2: 467 (1976)
BOBERG J., BOBERG M., GROSS R., GRUNDY S., AGUSTIN J., and BROWN W.V.
Atherosclerosis 27: 499 (1977)
EHNHOLM C., HUTTUNEN, J.K., KINNUNEN P.K.J., MIETTINEN T.A. and NIKKILA E.A.
New Engl. J. Med. 292: 1314 (1975)
AGUSTIN J., and GRETEN H.
Progress in Biochemical Pharmacology Lipoprotein Metabolism
Vol. 15 p-5 Ed. S. Eisenberg, Karger Basel. 1979
ZILVERSMIT D.B.
Circulation Res. 33: 633 (1973)
SIRTORI C.R., CATAPANO A., GHISELLI A., INNOCENTI G.C. and RODRIGUES J.L.
Atherosclerosis 26: 79 (1977)
RODRIGUES J.L., CATAPANO A., GHISELLI G.C., and SIRTORI C.R.
Adv. Exp. Med. Biol. 67: 169 (1977)
CONTI F., SIRTORI M., GIANFRANCESCHI G. TREMOLI E. and SIRTORI C.R.
Arteriosclerosis 1: 53 (1976)
SIRTORI C.R., TREMOLI E., SIRTORI M., CONTI M. and PAOLETTI R.
Atherosclerosis 26: 583 (1977)
TIENGO A., FEDELE D., NOSADINI R., and CREPALDI G.
International Conference on Atherosclerosis p-327. Ed.
Carlson, L.A., Raven Press. New York, 1978.
DESCOVIC G., MONTAGUTI U., CEREDI C. COCUZZA E., and SIRTORI C.R.
Artery 4:348 (1978)
HOFF H.F.,
Artery 4: 178 (1979)
HOLLANDER W.
Exp. Mol. Path. 25:106 (1976)
HOLLANDER W., PADDOCK J., and COLOMBO M.
Exp. Mol. Path. 30: 144 (1979)
GODSTEIN J.L., HO Y.K., and BROWN M.S.
Proc. Natl. Acad. Sci. U.S.A. 76:333 (1979)
FILIPOVIC I., SCHWARTZMANN G., MRAZ W., WIEGAND H. and BUDDECKE E.
Eur. J. Biochem. 93:51 (1979)
CAMEJO G., LOPEZ A., VEGAS H., and PAOLI H.
Atherosclerosis 21:77 (1975)
AVILA E.M., LOPEZ F., and CAMEJO G.
Artery 4: 36 (1978)
BERENSON G.S., SRINIVASAN S.R., DOLAN P.J. and RADHAKRISHNAMURTHY B.
Circulation 44 (II): 6 (1971)
IVERIUS P.H.,
J. Biol. Chem. 247:2607 (1972)
SMITH E.B.
Am. J. Pathol. 86:665 (1977)

DAY C.E., POWELL J.R. and LEVY R.S.
Artery 1: 126 (1975)
SIRTORI C.R., CATAPANO A., GHISELLI G.C. and MALINOW W. R.
Artery 2: 390 (1976)
GOLDSTEIN J.L., BASU S. K., BRUNCHEDE G.Y., and BROWN M.S.
Cell 7: 85 (1976)
SHIMAMOTO T.J.
Jap. Atheroscler. Soc. 1: 215 (1974)
NUMANO F., MAEZAWA H., SHIMAMOTO T. and ADACHI K.
Ann. N.Y. Acad. Sci. 231: 453 (1976)
WINDAUS A.Z.,
Physiol. Chem. 67:174 (1910)
KRITCHEVSKY D., TEPPER S. A., and KOTHARI H.V.
Artery 1: 437 (1975)

CLINICAL AND PATHOGENIC ASPECTS OF FREQUENT SYMPTOMATIC HYPERLIPOPROTEINEMIAS

R. Mordasini and W. Riesen

Lipid Research Laboratory and Institute for Clinical and Experimental Cancer Research (Tiefenauspital) of the University of Berne (Switzerland)

Secondary hyperlipoproteinemias are defined as disorders in lipid metabolism which occur in correlation with a different basic disease and which are reversible after successful treatment of this disease. Secondary hyperlipoproteinemias are frequent. Out of a total of 10-15% hyperlipoproteinemias in an adult "civilized"population about 40% can be defined as secondary forms.

During the last few years several new techniques have been developed which make it possible to study lipoprotein metabolism, namely also the catabolism of triglyceride-rich lipoproteins, in more detail. Consequently, the interest in the secondary forms of hyperlipoproteinemia has grown considerably, which is mainly due to the fact that the changes observed in secondary lipid disorders can serve as models for a better understanding of lipoprotein metabolism. The detection of secondary lipid disorders, however, is not only of scientific, but also of diagnostic and therapeutic interest. Above all, the exclusion of secondary hyperlipoproteinemias is important in order to avoid unnecessary, expensive, and - possibly - even harmful lipid-lowering drug treatment.

Three different pathogenic mechanisms play a major role in the development of secondary lipid disorders: (1) an increased lipoprotein synthesis, (2) disturbances in lipoprotein catabolism, and (3) the formation of abnormal lipoproteins. In the majority of the secondary hyperlipoproteinemias disturbances of the catabolism of triglycerid rich lipoproteins play the predominating pathogenic role.

The most frequent secondary lipid disorders are observed in obesity, alcohol consumption, diabetes mellitus, liver and renal diseases as well as in thyroid function disorders. Furthermore, lipoprotein alterations occur as a consequence of the application of certain drugs, especially diuretics contraceptives and corticosteroids. Opinions disagree on the question whether hyperlipoproteinemias in obesity and alcohol consumption belong to the secondary forms. According to our definition given in the introduction, however, they must be classified as secondary hyperlipoproteinemias.

Hyperlipoproteinemia in obesity

Hyperlipoproteinemia can be found in approximately 30% of obese persons, and the incidence in serious obesity is considerably higher. They usually show a pure or predominant hypertriglyceridemia with an increase in the mainly triglyceride-transporting lipoprotein class, the very low density lipoproteins (VLDL), i.e. a Type IV-pattern according to Fredrickson (5). It is mainly caused by an increased lipoprotein synthesis in the liver following hypercaloric nutrition.

In general, hyperlipoproteinemia in obesity can be normalized completely by adequate weight reduction (quantitative diet). Occasionally, additional qualitative dietary measures (abstinence of alcohol, carbohydrate restriction) become necessary. However, these measures are usually applied anyhow in the normal weight reduction diet (24).

Hyperlipoproteinemia in alcohol consumption

Alcohol increases the hepatic triglyceride synthesis (3). Nevertheless, regular alcohol consumption leads to hyperlipoproteinemia only in a small number of persons. As in obesity, it effects a Type IV hyperlipoproteinemia (elevation of triglycerides and VLDL). Therefore, it can be assumed that alcohol consumption "within ordinary limits" induces hyperlipoproteinemia only in persons with a "special" disposition. In alcoholics, however, a rise in blood lipid concentrations occur much more frequently. It is mainly caused, a part from the above-mentioned increased lipoprotein synthesis, by alcoholic hepatopathy and sometimes pancreatitis, which is often found in alcoholics. Marked hyperlipoproteinemias are especially observed in patients with Zieve syndrome and clinically manifest alcohol-induced pancreatitis. In both cases, patients usually show a Type V hyperlipoproteinemia (exogenous and endogenous hypertriglyceridemia with chylomicronemia and elevation of VLDL). Studies on the pathogenesis show a reduced post heparin lipolytic activity (PHLA, 10), i.e. a deficient clearing of triglyceride-rich lipoproteins in plasma due to a markedly reduced LPL-activity whereas HTGL-activity is found to be within normal limits (13). In persons with "normal" metabolism, regular alcohol consumption not only results in the above-mentioned increased triglyceride and VLDL synthesis in the liver. It is also followed by a considerable rise of the high density lipoproteins (HDL),whose protective effect against the development of atherosclerotic vascular complications could be demonstrated by large-scale epidemiological studies (12,16). With regard to the concomitant serious problems, however, the recommendation of aethanol as "prophylaxis of atherosclerosis" must certainly be considered with utmost caution and reserve!

The therapy of alcohol hyperlipoproteinemia consists in consequent abstinence of alcohol. Further dietary measures or even lipid-lowering drug treatment hardly ever become necessary.

Hyperlipoproteinemia in renal disease

Cardiovascular diseases are a most frequent and often fatal complication in patients under hemodialysis and after renal transplantation (19). In both groups of patients a high incidence of hyperlipoproteinemias as a major risk factor of coronary heart disease can be found (in more than 50% of the cases). This is also true in patients with conservatively treated severe renal insufficiency (11, 19). In the nephrotic syndrome, a secondary increase of blood lipids is obligatory. In patients under hemodialysis and with conservatively treated cronic uremia, the predominant pattern is Type IV hyperlipoproteinemia. Patients after renal transplantation more often show a Type IIa (pure hypercholesterolemia with LDL increase) and IIb pattern (predominant hypercholesterolemia with LDL increase and concomitant hypertriglyceridemia). Slight cases of the nephrotic syndrome mostly induce pure hypercholesterolemia, whereas in severe cases we usually find a Type IIb pattern (Table I).

Regarding pathogenesis a selective enzyme deficiency of hepatic triglyceride lipase (HTGL) can be observed in the uremic hyperlipoproteinemia. This enzyme deficiency leads to an accumulation of abnormal triglyceride-rich LDL and of VLDL (19). Apart from that, there are signs for an increased hepatic lipoprotein synthesis, which, however, is most probably not an essential factor in this metabolic situation (11).

Hemodialysis does not substantially alter the uremic lipid disorder, although the choice of the dialysis procedure is believed to have a certain effect on the degree of hyperlipoproteinemia (21).

Hypercholesterolemia in the nephrotic syndrome shows a strong inverse correlation to serum albumin concentration. In severe cases,

Table I. Dyslipoproteinemias in chronic renal disease

	chronic uremia	hemodialysis	transplantation	nephrosis
predominant type of HLP	IV	IV	II b	IIa / IIb
frequency ~	20-40%	30-60%	90%	100%
atherogeneity LDL HDL	++ ↑ ⬇	++ ↑ ⬇	+++ ⬆ ⬇	+ ⬆ ⬆
pathogenic mechanism	HTGL-activity ⬇	HTGL-activity ⬇	Corticosteroid treatment	LP-synthesis ⬆ LPL-activity ⬇

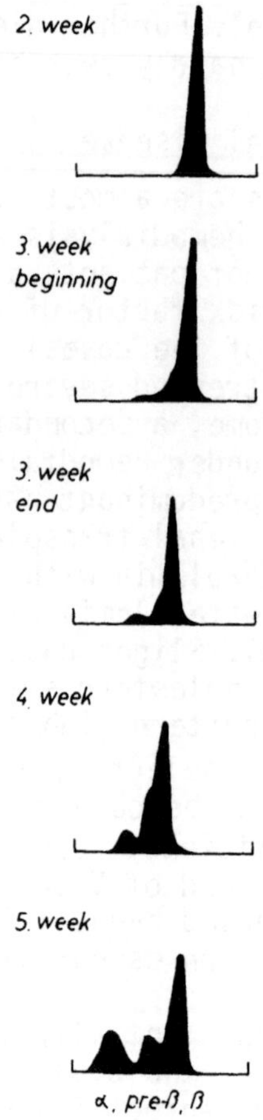

Figure I. Changes in lipoprotein electrophoresis in the course of acute hepatitis. Densitometric evaluation of lipoprotein electrophoresis on agarose with lipid staining.

hypertriglyceridemia occurs in addition to the cholesterol increase; it is probably closely related to a selective deficiency of lipoprotein lipase (LPL, 20). In the nephrotic syndrome, the HDL levels are generally increased (20); this is contrary to the markedly decreased HDL concentrations found in the other secondary hyperlipoproteinemias in chronic renal diseases (2,19). The high HDL-levels in the nephrotic syndrome could recently been confirmed in our laboratory by the measurement of the major HDL-apoproteins A1 and A2, which were found to be significantly increased in patients with the nephrotic syndrome

Table II. Dyslopoproteinemias in liver disease

	acute hepatitis	chronic hepatitis	cirrhosis	cholestasis
predominant type of HLP	IV atypical	II b atypical	hypolipo-proteinemia	II b, atypical Lp-X
frequency ~	90 %	50 %	90 %	90 %
pathogenic mechanism	HTGL activity ↓ disturbed synthesis of Apo $A_I + A_2$	HTGL activity ↓	LP-synthesis ↓	?

as compared to healthy controls. These elevated HDL-levels in the nephrotic syndrome are probably, as well as the increase in LDL, the result of an increased lipoprotein synthesis on the whole.

Contrary to conservatively treated uremics and those under hemodialysis, no enzyme deficiency could be detected in patients after renal transplantation. Hyperlipoproteinemia in this group of patients might be mainly due to corticosteroid medication obligatory after renal transplantation. In patients who are undergoing long-term corticosteroid treatment for other reasons (for example primary chronic polyarthritis), similar secondary lipid disorders can be observed (1).

Regarding the accumulation of cardiovascular complications, the question of a lipid-lowering drug treatment of hyperlipoproteinemia following uremia and renal transplantation must be considered. This is of special importance as hyperlipoproteinemia in these patients can hardly be influenced by dietary measures. So far, however, no large-scale studies on lipid-lowering drug treatment in chronic renal disease have been reported; therefore this question must still remain unanswered.

Alterations of serum lipoproteins in liver disease

Due to the central function of the liver in lipoprotein metabolism, diseases affecting this organ lead to marked dyslipoproteinemias (Table II).

Acute hepatitis is usually followed by marked hypertriglyceridemia with a concomitant slight increase in plasma cholesterol and phospholipids (18,22). Hypertriglyceridemia reaches its peak during the second week, provided an uncomplicated course of the disease. Striking and characteristic changes can be demonstrated in lipoprotein electrophoresis (with the usual lipid staining): at the end of the first week of hepatitis, the "disappearance" or far-reaching reduction of the α-band with a simultaneous amalgamation of the pre-β and β-lipoprotein bands can be observed (fig.1). The changes in lipoprotein electrophoresis are normalized parallel to the improvement of the chemical lipid status, commonly during the fourth week of the disease (18).

As a possible cause of hypertriglyceridemia in acute hepatitis a transitory deficiency of hepatic triglyceride lipase (HTGL) could recently be detected (14). The electrophoretic changes are probably due to a disturbance in the synthesis of the apolipoproteins A1 and A2, a decrease in the transport capacity of the HDL for triglycerides (which instead largely bind bile acids), and to the accumulation of IDL resulting from disturbed VLDL catabolism (15).

In chronic hepatitis the alterations of serum lipids and lipoproteins are dependent on the activity of the inflammation: during the acute stage, all serum lipids are slightly elevated and lipoprotein electrophoresis shows an extensive amalgamation of pre-β- and β-lipoprotein bands. In the interval changes are scarcely characteristic.

A considerable reduction of the amount of functioning liver cells (cirrhosis of the liver) results in hypolipoproteinemia with a predominant decrease of β-lipoproteins (LDL). This hypobetalipoproteinemia is caused by diminished lipoprotein synthesis in the affected liver.

Intra-and extrahepatic cholesttasis leads, independently of his etiology, to the incidence of an abnormal lipoprotein specific for cholestasis, lipoprotein-X (Lp-X). Lp-X is an abnormal β -lipoprotein with an extremely low protein moiety (approximately 6% with 40% of albumin), and it is characterized by a high content of phospholipids and esterified cholesterol. This abnormal β-lipoprotein, which could, apart from cholestatic processes, so far only be observed in an extremely rare familial disease (LCAT-deficiency, lecithin-cholesterol-acyltransferase deficiency, 18) is easili shown in migration electrophoresis due to its characteristic electrophoretic behaviour. Contrary to all other lipoproteins, the Lp-X migrates to the cathode. For the detection of Lp-X simple tests are commercially available.

Despite the high sensitivity, simplicity and good specificity of this Lp-X-test, it was not generally accepted as a method for the exclusion of cholestasis. This is mainly due to the fact that Lp-X can be missing in certain cases despite undoubtedly established cholestasis (17). It seems, However, that methodological improvements of the Lp-X test can solve this problem.

Diabetes mellitus and hyperlipoproteinemia

Hyperlipoproteinemia is found in approximately 40% of patients with diabetes mellitus. Regarding the frequency of cardiovascular complications in diabetics, the increase of blood lipids in this group of patients deserves special attention. A distinction must be made between secondary hyperlipoproteinemia resulting from the special metabolic situation in diabetes and the primary forms which occur parallel to diabetes mellitus. For practical purposes (23), they can be distinguished according to their reaction to the therapy of diabetes - whether they can be normalized by "optimal" control of the disease (secondary hyperlipoproteinemia) or whether they persist unchanged (primary lipid disorder). More detailed subdivisions which have been made to some extent seem to be of little use with regard to the present lack of diagnostic and therapeutic consequences.

Primary and secondary hyperlipoptoteinemias in diabetes mellitus principally belong to the type IV pattern according to Fredrickson (5) Type IIb and type V hyperlipoproteinemias are also frequently observed in this disease.

Investigations on the pathogenesis of diabetic hyperlipoproteinemia are still in their early stages. An important factor is the increase of hepatic VLDL-synthesis, apart from probably less important disturbances in lipoprotein catabolism (reduced PHLA due to LPL deficiency) which have recently been reported (23,24).

The high incidence of atherosclerotic vascular complications in diabetes in any case requires the therapy of hyperlipoproteinemia as a major risk factor of coronary heart disease. Consequent dietary treatment of diabetes mellitus also meets to a large extent the requirements of dietary treatment of the predominant type IV hyperlipoproteinemia. The main objective is a weight reduction to abnormal even better ideal weight, a drastic reduction of the intake of easily absorbable carbohydrates and a strict abstinence of alcohol (24). When controls in short intervals confirm adherence to dietary treatment, whereas lipid levels continue to be elevated, addicional lipid-lowering drug treatment must be taken into consideration.

Dyslipoproteinemias in thyroid diseases.

In hyper-as well as in hypothyreosis secondary changes in serum lipoproteins are considered the rule. Hyperthyreosis is usually followed by a more or less marked hypocholesterolemia, in hypothyreosis type IIa or type IIb hyperlipoproteinemia are almost regularly observed.

In the pathogenesis of hyperlipoproteinemia in hypothyreosis delayed LDL catabolism is the predominant mechanism. It predominates in its effect on plasma LDL levels over a reduced LDL synthesis in this metabolic situation. Moreover, an abnormal lipoprotein (β-HDL) is often found in hypothyreosis (25). However, no studies on its clinical relevance have been reported yet. In hyperthyreosis, a highly accelerated LDL catabolism with increased secretion of bile salts in the stool predominates over an increased cholesterol synthesis (25).

The secondary lipid disorders in hypo-and hyperthyreosis are rapidly normalized by adequate treatment of the thyroid function disorder. Observation of the lipid status, therefore, can be regarded as a rather reliable parameter for the control of therapy.

Lipid disorders in the application of drugs

Serum lipoprotein concentrations are affected by numerous drugs.

The secondary increase of blood lipids due to the long-term administration of corticosteroids has been known for several years (1). Generally, it induces type IIa or type IIb hyperlipoproteinemia. The extent of this lipid disorder seems to be dependent on the corticosteroid dosage. A complexe multifactorial group of etiologies have been noted in the evaluation of secondary hyperlipoproteinemia in women receiving mixed estrogen-progestin oral contraceptives containing both estrogens and progestins have generally been found to induce mild increments in plasma triglycerides and less substantial increments in plasma cholesterol (4,6). These changes appear to be caused by the estrogen component, as does the depression of post--heparin lipolytic activity (PHLA), deterioration of glucose tolerance and elevation of plasma insulin and growth hormone. The progestational component appears to have little to no effects on plasma lipids and may lower triglycerides in normals and patients. When oral contraceptives are taken in a previously existing hyperlipoproteinemia, they may lead to an important increase of serum triglycerides. In persons with a corresponding disposition (familial incidence of hyperlipoproteinemia!), a relevant elevation of blood lipids should be excluded before the administration of oral contraceptives.

Diuretic treatment may also interfer with lipid metabolism (7). Several investigators noted a tendency for increased serum trigliceride and/or total cholesterol levels in patients treated with variuos diuretics. However, information on serum lipoproteins which are more important correlates of atherogenicity than lipid levels per se has been limited to a few studies. Thus, chlorthalidone and furosemide were found to elevate the serum LDL-fraction (7).According to more recent results in our laboratory there are important differences between the individual diuretics concerning their effect on the serum lipoproteins. Up to now, however, no well documented studies have been published considering the question in how far hyperlipoproteinemia resulting from diuretic therapy is significant as a risk factor of coronary heart disease and therefore might question the success of long-term diuretic treatment in hypertension. Several other drugs are likely to effect secondary changes in lipoprotein metabolism as well. At present, however, no large scale studies exist on this topic. In this connection, not only serum lipoprotein concentrations deserve attention. Among others, also questions regarding changes in the bile composition in the sense of increased lithogenicity are of growing importance!

The above-mentioned frequent secondary lipid disorders may indicate which marked dyslipoproteinemias can be induced by most different basic diseases. The type and degree of these secondary disorders of lipid metabolism may give interesting diagnostic and therapeutic information. Most important, however, and this must be stressed again in these final remarks - is a thorough screening for secondary forms of hyperlipoproteinemia before any lipid-lowering for secondary forms of hyperlipoproteinemia before any lipid--lowering treatment, especially previous to drug therapy. By means of this screening many unnecessary, expensive and perhaps even harmful lipidlowering drug therapies can be avoided.

REFERENCES:

1. Bagdade J., Porte D., Bierman E.: Steroid induced lipemia. A complication of high-dosage corticosteroid therapy. Arch. Int. Med. 125, 129 (1970).
2. Bagadade J., Albers J.: Olasma high-density lipoprotein concentrations in chronic-hemodialysis and renal-transplant patients. New Engl. J. Med. 296, 1436-1439 (1977).
3. Baraona R., Lieber C.: Alcoholic hyperlipemia. Handbuch der inneren Medizin, 5. Auflage, Band 7, 389. Ed. G. Schettler et al. Verlag Springer Berlin, Heidelberg, New York (1976).
4. Corredor D.G., Mendelson L.V., Sabeh G., Sunder J.H., Danowski T.S.: Serum lipids during oral contraceptives exposure. Clin. Pharmacol. Ther. 11, 188 (1970).
5. Fredrickson D., Levy R., Lees R.: Fat transport in lipoproteins - an integrated approach to mechanisms and disorders. New Engl. J. Med. 276, 34, 94, 148, 215, 273 (1967).
6. Glueck C.J., Scheel D., Fishback J., Steiner P.:Progestagens, anabolic-androgenic compounds, estrogens: Effects on triglycerides and post-heparin lipolytic enzymes. Lipids 7, 110 (1972).
7. Gluck Z., Weidmann P., Mordasini R., Bachmann C., Riesen W., Peheim E., Keusch G., Meier A.: Increased serum low-density lipoprotein cholesterol in men treated short-term with the diuretic chorthalidone. Metabolism 29, 240 (1980).
8. Gjone E., Javitt N.B., Blomhoff J.P., Fausa O.: Studies of lipoprotein-X (Lp-Y) and bile acids in familial LCAT deficiency. Acta med. Scand. 194, 377-378 (1973).
9. Godstein J.L., Brown M.S.: Binding and degradation of low density lipoproteins by cultured human fibrolast: Comparison of cells from a normal subject and from a patient with homozygous familial hypercholesterolemia. J. Biol. Chem. 249, 5153-5162 (1974).
10. Greten H., De Grella R., Klose G., Rascher W., de Gennes J., Gjone E.: Measurement of two plasma triglyceride lipases by an immunochemical method: Studies inpatient with hypertriglyceridemia. J. Lipid Res. 17, 203-210 (1976).
11. Ibels L., Simons L., King J.: Studies on the nature and causes of hyperlipoproteinemia in uraemia, maintenance dialysis and renal transplantation. Quart. J. Med. 44, 601 (1975).
12. Johansson B., Nilsson P.: Alcohol comsumption and high-

-density lipoprotein.
New Engl. J. Med. 298, 633 (1978).

13. Kaffarnik H., Schneider J., Mordasini R., Liesenfeld A., Muhlfellner G., Schubotz R.: Behaviour of lipoprotein fractions, lipoprotein lipase and hepatic triglyceride lipase during short-term and long-term uptake of ethanol in healthy subjects.
Submitted to: Atherosclerosis.
14. Klose G., Windelband J., Weizel A., Greten H.: Secondary hypertriglyceridaemia in patients with parenchymal liver disease.
Europ. J. Clin. Invest. 7, 557-562 (1977).
15. Middelhoff G., Mordasini R., Stiehl A., Greten H.: A bile acid high density lipoprotein (HDL) in a acute hepatitis.
Scand. J. Gastroent. 14-267 (1979).
16. Miller N., Miller G.: High-density-lipoprotein and atherosclerosis.
Lancet I, 1033 (1975).
17. Mordasini R.C., Berthold S., Schlumpf E., Keller H., Riva G.: Lipoprotein-X bei hepatobiliaren Erkrankungen.
Schweiz. med. wschr. 105, 863-872 (1975).
18. Mordasini R., Berthold S., Schlumpf E., Riva G.: Veranderungen der Serumlipide und-lipoproteine bei akuter hepatitis.
Schweiz. med. Wschr. 106, 1173-1182 (1976).
19. Mordasini R., Frey F., Flury W., Klose G., Greten H.: Selective deficiency of hepatic triglyceride lipase in uremic patients.
New Eng. J. Med. 297, 1362 (1977).
20. Oetliker O., Mordasini R., Lutschg J., Riesen W.:Lipoprotein metabolism in nephrotic syndrome in chilhood.
Ped. research 14, 64 (1980).
21. Panzella G., Castellani A., Mioni G., Cristinelli Cecchettin M., Mingozzi G., Majorca R.: The worsening of glucose tolerance after a single dialysis: a factor promoting hyper-triglyceridemia?
VI. Int. Congress of Nephrology, Firenze (1975) Seite 950.
22. Seidel D.: Hyperlipoproteinamie bei Erkrakungen der leber.
Schweiz. med. wschr. 105, 857-862 (1975).
23. Schlierf G.: Hyperlipidamien (Hyperlipoproteinamien) bei Diabetes mellitus, Handbuch der inneren Medizin, 5. Auflage, Band 7, 363 Ed. G. Schettler et al.
Verlag Springer Berlin, Heidelberg, New York (1976).
24. Schlierf G., Oster P.: Diagnostik und therapie der Fettstoffwechselstorungen. Thieme Verlag Stuttgart (1978).
25. Wieland H.: Einfluss der Schilddruse auf den lipidstoffwechsel. Hanbuch der inneren Medizin, 5. Auflage, Band 7, 423. Ed. G. Schettler et al. Verlag Springer Berlin, Heidelberg, New York (1976).

CRITICAL COMPARISON OF COMMERCIALLY AVAILABLE METHODS FOR THE DETERMINATION OF HDL-CHOLESTEROL

C. Luley, A. Lippert, and W. Schwartzkopff

Metabolic Disease Clinic in the Department of Medicine at the Klinikum Charlottenburg of the Free University Berlin

SUMMARY

Three commercially available methods were compared in respect to correlation and precision.

1. Precipitation with phosphotungstate acid ("PWS")*,
2. Electrophoresis with subsequent precipitation in agarose ("electrophoresis")**,
3. Precipitation of Apo-B-containing Lipoproteins with Concanavaline A ("ConA")***.

As a standard we used ultracentrifugation at a density of 1.006 g/ml with sudsequent precipitation of LDL-lipoproteins using dextranesulphate (1) (" modified UZ ").

Mean values for pre-β-cholesterol were 4.7 mg/dl less than UZ values and triglyzerides/5 values were 15.6 mg/dl higher. For HDL--cholesterol ConA produced results similar to those with modified UZ, while HDL-cholesterol (PWS) and α-cholesterol (electrophoresis) were lower by 4.2 mg/dl and 5.5 mg/dl, respectively. We found good correlations among all methods in LDL-cholesterol determinations. There was good agreement between PWS and modified UZ, while electrophoresis provided results 13.6 mg/dl higher and ConA 22.1 mg/dl lower than the standard.

INTRODUCTION

HDL-cholesterol is regarded as an important indicator of coronary risk. Therefore there is wide interest in determining concentrations of this lipoprotein as a routine parameter. As a result a number of laboratory kits for the determination of HDL-cholesterol has

---

* Boehringer, Germany ** Immuno, Austria
*** Fournier, France

appeared on the market. The purpose of our paper is to answer the question to what extent results obtained with these different methods are comparable, especially in respect to correlation and precision.

In addition we determined the proportion of lipids in each of the fractions of the UZ and found to what extent these proportions vary among type II a, II b and IV hyperlipoproteinemias. This has importance for quantification of lipoprotein cholesterol in electrophoresis since this method uses stable factors for the calculation, regardless or the type of hyperlipoproteinemia.

METHODS

We measured HDL-cholesterol in 116 samples taken from patients in our metabolic disease service. Using the commercial methods we followed the manufacturers instructions.

UZ: 6 ml serum were centrifuged 25 hours at 105.000 g at 4º C. To 2 ml of the infranatant 100/ul of a solution of 2 N dextrane--sulphate 500 (Pharmacia, Sweden) and 100/ul of a solution of 20 g/l $MgCl_2$ were added. The solution remains 20 min. at 4º C for complete reaction. After centrifugation (20 min., 1500 g) HDL-cholesterol is determined with an enzymatic test.

Cholesterol, triglycerides and phospholipids in the UZ-fractions were determined by means of laboratory kits provided by Boehringer, Mannheim.

STATISTICAL ANALYSIS

The calculation was done on a Cyber 175 using SPSS in version 8. The regression line is described by the " Borner " equation; the U-test (Man-Whittney-Test) was used for calculation of statistical significance.

RESULTS AND DISCUSSION

The coefficients of variation (Tab. 1) do not vary in an unusual way 2. While the 3 methods give results in the same range,

Table 1. Coefficents of variation of all methods

	VLDL-CH prae-β-CH	LDL-CH β-CH LpB	HDL-CH α -CH LpA
PWS	-	3,9 %	5,1 %
mod. UZ	8,1 %	4,6 %	6,1 %
Electrophoresis	5,2 %	4,5 %	4,8 %
ConA	-	3,2 %	4,8 %

the CV of the modified UZ is slightly higher.

Fig. 1 shows the first comparison: VLDL-cholesterol determined by ultracentrifugation compared with pre-β-cholesterol and Tg/5, as in the Friedewald Equation. As can be seen in the graphs, the mean values and the formulae of the regression-lines, Tg/5 gives the highest and electrophoresis the lowest values.

In a comparison of HDL-cholesterol-Values (fig 2), the four methods correlated fairly well-between 0.91 and 0.97. In general we found higher values with the modified UZ. It was observed in separate mearurements that the dextrane sulphate solution itself produces an increase in the subsequent determination of cholesterol, a phemomenon which has been reported by other authors 3,4 .

The greatest differences were seen in the comparison of LDL--cholesterol values. The cholesterol values - which were measured after ConA-precipitation - could not be compared directly with those of the other methods, since they show cholesterol not only of the LDL-lipoprotein, but of all lipoproteins ApoB. We corrected by subtracting Tg/5. After this correction a relatively high correla-(0:97-098) resulted. In comparison to the modified UZ, electrophoresis produced the highest values, followed by PWS, while there was a considerable difference between modified UZ and ConA. A possible explanation for this difference could be the Tg/5 values used for correcting both PWS and ConA which is probably too high a correction factor. Another reason for the lower values with ConA is a probable interaction of the reagents with the subsequent cholesterol determination; we were able to demonstrate this in separate experiments in which the values obtained were about 6% lower.

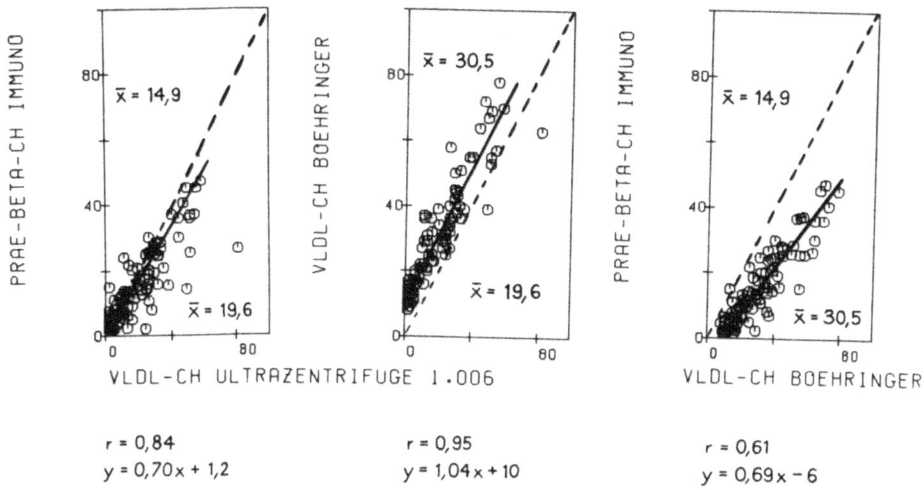

Figure 1. Mean values, correlation factors, regression lines and equation for the comparison of VLDL-cholesterol.

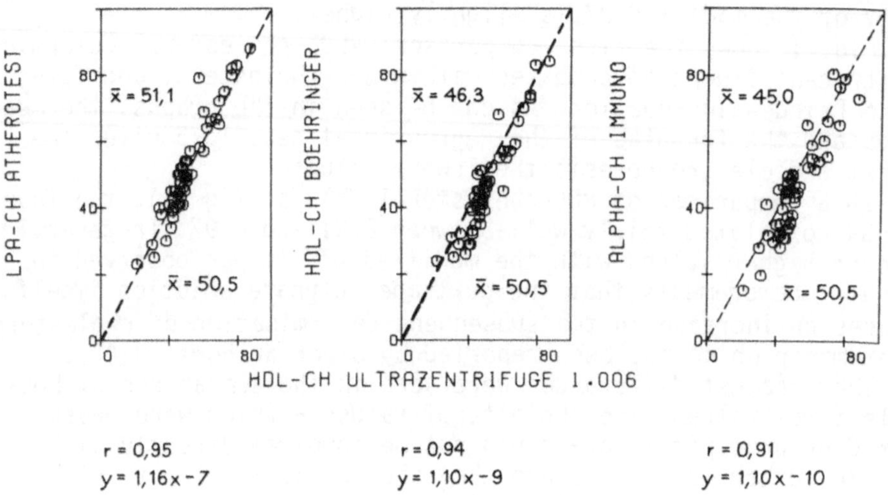

Figure 2. Mean values, correlation factors, regression lines and equations for the comparison of HDL-cholesterol measured by either method.

In addition to the report comparison we investigated a problem which arises in connection with electrophoresis. The calculation is done using factors which are based on the cholesterol concentration of each lipoprotein fraction. e.g. 15% pre-β-lipoprotein, 45% in β-lipoprotein and 18% in alpha-lipoprotein . We wanted to check whether this proportion of cholesterol is stable: (1.) in either fraction and (2.) regardless of the type of HLP.

In order to test the assumption of fixed proportion we measured the triglycerides and phospholipides in addition to cholesterol in each of the ultracentrifugation fractions. It was then possible to calculate the total lipids and to express the proportion of each lipid as a Percentage of the total lipids in either fraction. Putting the lipid proportions in an order of the common hyperlipoproteinemia types we compared them with the healthy collective. As is indicated by* and *** (Tab. 2) it is evident that these proportions do change significantly, if a patient suffers from hyperlipoproteinemia, especially typy II b and IV.

It could now be expected that choleterol values calculated using the above mentioned factors would be somewhat high in hypertriglyceridemia.

Tab. 3 shows that the results obtained with the three methods do not differ greatly despite the varying proportions of cholesterol

Table 2. Proportions (%) of lipids in the fractions of the modified UZ in type IIa, IIb and IV

	VLDL			LDL		
	TG	CH	PH	TG	CH	PH
Healthy	60	12	28	13	49	38
II a	61	15*	24	11*	52*	37
II b	66*	17***	17***	14*	49	37
IV	66***	16***	17***	17***	44***	39

	HDL		
	TG	CH	PH
Healthy	7	31	62
II a	8	31	61
II b	13***	26***	61
IV	12***	26***	62

*=p 0,05 ***=0,0001

Table 3. Mean cholesterol values as measured with the different methods in type II a, II b and IV

	LDL-cholesterol (mg/dl)				HDL-cholesterol(mg/dl)			
	elpho.	PWS	ConA	mod. UZ	elpho.	PWS	ConA	mod.UZ
Heathy	160	149	130	150	50	54	57	52
II a	245	231	207	231	46	51	50	47
II b	257	239	213	232	35	36	40	35
LV	160	135	119	143	34	36	37	37

found in the UZ fractions (in the three types of hyperlipoproteinemia disease).

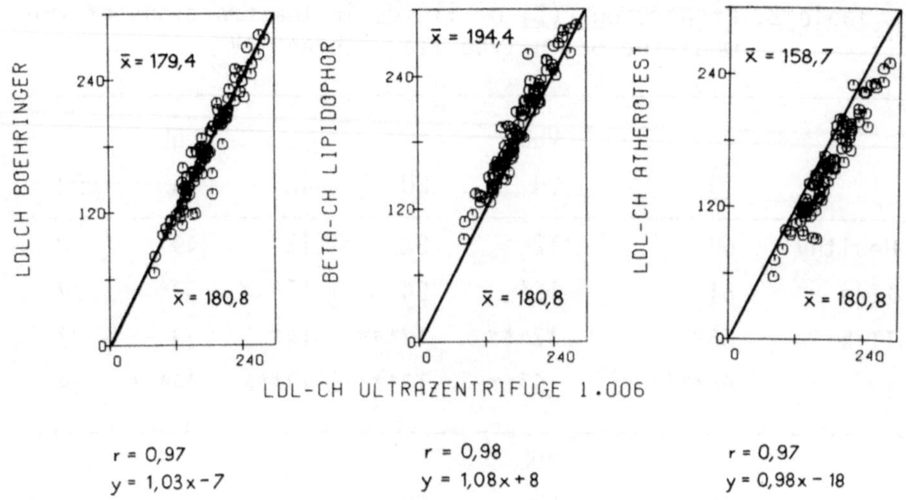

Figure 3. Mean values, correlation factors, regression lines and equations for the comparison of LDL-cholesterol measured by either method.

REFERENCES

1. Assmann, G., Simantke, O., Schaefer, H.E. Smootz, E. J. Clin. Invest. 60:1025-1035 (1977)
2. Wieland, H., Niazi, M., Barthlomé, M., Seidel, D. Arztl. Lab. 26:257-266 (1980)
3. Draeger, B., Wahlefeld, A., Lipoproteine und Herzinfarkt publ. by H. Greten, P.D. Lang, G. Schettler, Gerhard Witzstock, Baden-Baden, Koln, New York (1979)
4. Warnick, G.R., Cheung, M.c., Albers, J.J. Clin. Chem. 25/4: 596-604 (1979).

BILE ACID METABOLISM IN HYPERLIPOPROTEINEMIAS

Tatu A. Miettinen

Second Department of Medicine, University of Helsinki

00290 Helsinki 29, Finland

INTRODUCTION

Cholesterol is eliminated from the body mainly via the gastrointestinal tract into feces as cholesterol itself or after hepatic conversion to cholic acid and chenodeoxycholic acid as these primary bile acids or as their bacterial transformation products. The magnitude by which cholesterol elimination takes place as bile acids can be measured by quantitation of fecal bile acids or by the isotope dilution technique. In the present paper some parameters of bile acid metabolism are dealt with in different hyperlipoproteinemias mainly on the basis of our own experience on fecal bile acid measurements. The capacity to produce bile acids has been tested after administration of cholestyramine. In a recent review a computerized analysis has been performed on reported fecal steroids of individual subjects with different hyperlipoproteinemias.[1]

Primary hyperlipidemias

Earlier studies showed that in a family with hypercholesterolemia and tendon xanthomata bile acid production was fairly low and was negatively correlated with the serum cholesterol level.[2] Subsequent studies on a fairly large number of subjects with heterozygous xanthomatotic familial hypercholesterolemia actually revealed that the mean bile acid synthesis was subnormal (Fig.1).[3] Furthermore, an association of obesity or hypertriglyceridemia, two factors frequently enchancing bile acid formation, caused no increase in bile acids of those patients (Table I). The reason for the subnormal bile acid production is not apparent at the moment. Sterol balance data indicate that the overall synthesis of cholesterol is within the low normal limits (Table II). The hepatic synthesis may even be subnormal.[4] Thus, the synthesis of bile acids

Table I. Serum lipids and fecal bile acids in patients with familial hypercholesterolemia subgrouped according the obesity and hypertriglyceridemia

Group	No	Relative weight	Serum lipids, mmol/l Triglycer.	Cholesterol	Fecal bile acids mg/day	mg/kg/day
All subjects	69	1.22±0.02	1.49±0.07	11.9±0.4	176±9	2.7±0.1
Non obese II A	42	1.06±0.02	1.28±0.08	11.6±0.5	175±10	2.8±0.1
Non obese II B	12	1.06±0.03	2.27±0.10	13.3±0.8	142±21	2.4±0.3
Obese II A	11	1.32±0.04	1.23±0.13	12.1±0.4	172±24	2.6±0.2
Obese II B	4	1.39±0.10	2.47±0.12	11.3±1.1	280±17	2.9±0.1
Obese	15	1.34±0.04	1.56±0.19	11.9±0.3	201±22	2.7±0.2
II B	16	1.14±0.05	2.32±0.10	12.8±0.7	176±22	2.5±0.2

from newly formed cholesterol may also be low and could contribute to the subnormal bile acid values. The data in Table II indicate that the bile acid and cholesterol synthesis are interrelated.

The increase in bile acid synthesis after cholestyramine was lower than in a somewhat younger control group (Table II).[5] A closer analysis showed that when the hypercholesterolemic subjects were subgrouped according to increasing age by ten-year periods (youngest age group of 27 years matched for controls) the base line bile acid values were partly age-dependent and were about 1 mg/kg/day lower than in the controls (fig.1). On cholestyramine, however, the youngest age-group increased its bile acid excretion more than the other subgroups but the increase was less than in the age-matched control group. However, the few controls and patients with similar initial bile acid outputs appeared to increase their bile acid production equally. This is actually shown in Table II for the parents of the second family. Thus, under metabolic ward conditions on a low cholesterol standard diet the average base line bile acid excretion and the mean response to cholestyramine are lower in the heterozygous patients with xanthomatotic hypercholesterolemia than in normocholesterolemic subjects. Although the base line cholesterol synthesis is within the low control limits, the response to cholestyramine is subnormal (Fig. 1 and Table II).

Low fecal bile acid output could be associated with reduced biliary secretion of bile acids. Direct secretion measurements, however, revealed that, if anything, biliary bile acid secretion is higher than in the controls (Table III). Accordingly, reabsorption of bile acids should be effective and could contribute to the low

Table II. Serum lipids, fecal bile acids and cholesterol synthesis before and after cholestyramine treatment in patients with different types of hyperlipoproteinemias.

Type	No.	Serum lipids, mmol/l Cholesterol	Triglycerides	Fecal bile acids Total mg/kg/day	Cheno* % of total	Cholesterol synthesis mg/kg/day
Normal	6	5.2±0.4	0.86±0.09	4.1±0.5	39	14.1±2.3
		3.7±0.2	0.98±0.14	39.5±3.7	35	51.3±5.3
IIA X	12	11.2±0.5	1.20±0.11	2.7±0.3	49	11.3±1.2
		8.4±0.6	1.15±0.11	17.5±1.5	40	26.0±2.0
IIB X	6	12.1±1.4	3.04±0.67	2.9±0.3	46	12.2±1.8
		8.0±0.9	2.95±1.20	16.5±3.9	39	26.9±6.1
IIAXH	1	30.9	2.06	1.7	76	17.8
		31.1	2.00	35.5	37	53.7
IIAXM	1	9.9	0.80	1.9	44	9.3
		6.4	0.73	12.1	35	19.3
IIAXF	1	11.5	1.60	2.6	46	8.4
		8.9	2.14	17.9	55	24.7
IIAXH	1	26.7	1.00	7.1	43	25.3
		24.4	1.52	69.1	40	95.3
IIAXM	1	10.6	1.11	4.7	45	14.2
		5.9	1.19	37.6	39	53.3
IIAXF	1	8.7	0.92	3.7	-	13.3
		6.8	0.98	43.1	-	50.2
II B	7	7.9	3.47±0.48	3.8±0.5	42	13.5±1.6
		6.3	3.67±0.78	18.1±2.3	38	26.9±3.4
IV	7	7.0	4.90±1.27	4.7±0.5	41	17.2±2.2
		6.0	6.54±1.67	21.6±1.5	42	31.9±3.0

X = tendon xanthomata, H = homozygous, M = mother, F = father,
*Cheno include bile acids from chenodeoxycholic acid origin.
Cholestyramine 32 g/day (in homozygous children 24 g/day) for 10-12 days.
Cholesterol synthesis is a difference between fecal total steroids (bile acids plus sterols of cholesterol origin) and dietary cholesterol.

bile acid production provided that the bile acids inhibit cholesterol and/or their own synthesis in the liver.

In homozygous subjects with familial hypercholesterolemia the base line bile acid output varies from low to high. Results in Table II suggest that the child of the parents with a low output also exhibits a low value while the synthesis is high if the parents produce much bile acids. The overall cholesterol synthesis, however, appears to be higher in the children with double dose of hypercholesterolemic gen than in the parents with a single dose. In addition, the increase in bile acids and cholesterol synthesis during cholestyramine treatment is in the children about twice that in the parents. The plot of the initial bile acid or cholesterol synthesis against the increase caused by cholestyramine in the respective production rates shows a significant correlation (r = 0.859 and 0.970, respectively).

A combination of the present results from the two homozygous children with six cholestyramine-treated homozygous subjects from the literature (four from ref. 6 and two from ref. 7) showed that the mean base line bile acid output was significantly higher than in somewhat older heterozygous adults of Table II and tends to be higher than in normal adult controls. The overall baseline cholesterol synthesis was less significantly elevated. Furthermore, the increase in bile acids after the cholestyramine treatment was clearly above that in the heterozygous controls and slightly exceeded even that in the young normal adults of Table II. In addition, the increase in the synthesis of both bile acids and cholesterol was significantly correlated with the respective baseline production rates (r-values 0.865 and 0.970, respectively). It can be summarized that, though no age-matched controls are available, in general the baseline bile acid and cholesterol synthesis and the capacity to increase the synthesis are increased in homozygous subjects. The finding that in two homozygotes[8] of one family the baseline bile acid and cholesterol synthesis were low indicate that the entity of familial hypercholesterolemia is not homogenous.

A closer analysis of fecal bile acids in small subgroups of the control and hypercholesterolemic subjects showed that the low bile acid output was predominantly caused by impaired cholic acid production in familial hypercholesterolemia (Table II). Similar results have been obtained with the isotope dilution Technique.[9] Cholic acid synthesis appears to be related to the cholesterol production, newly formed cholesterol being preferentially utilized for the bile acid--especially cholic acid - synthesis. In fact an increase of cholesterol synthesis by cholestyramine was associated with a relatively high increase in the cholic acid production especially in hypercholesterolemic subjects with the low base line cholesterol and cholic acid synthesis. The relative amount of cholic acid in terms of percent of total fecal bile acids actually correlated positively with the overall cholesterol synthesis in Table II.

Patients with type II B pattern, genetically belonging to families with combined hyperlipoproteinemias, also have a low normal bile acid production under basal conditions (Table III).[5] The out-

Table III. Biliary steroid secretion in patients with different hyperlipoproteinemias

Group	Number of subjects	Steroid, mg/day	
		Cholesterol	Bile acids
Normal	7	837±105	8175±815
Type II A	7	895±113	10854±1790
Type IV	4	1268±224	8467±2111
PBC	10	435±65	3030±345
PBC*	10	307±25	112±36

PBC = primary biliary cirrhosis

*fecal steroids of endogenous origin

put of cholic acid tends to be subnormal that of chenodeoxycholate is normal. Kinetic bile acid studies have revealed similar changes except that the chenodeoxycholic acid turnover in terms of mg/day is significantly increased.[9] Administration of cholestyramine increases the bile acid and cholesterol production subnormally the relative increase of cholic acid tending to be higher than that of chenodeoxycholic acid (Table II).[5]

Many but not all of the patients with the type IV hyperlipoproteinemia appear to have increased fecal bile acid and neutral sterol output even when the obesity is taken into consideration.[5,9,10] Thus, cholesterol synthesis is frequently increased. Since fecal bile acids of hypertriglyceridemic subjects exhibit a negative correlation with serum cholesterol[5] it is apparent that especially those type IV subjects whose low density lipoprotein is low have a highest bile acid synthesis. The primary type IV hyperlipoproteinemia is actually an entity including subjects from type IV families, multiple type hyperlipoproteinemias and less characterized polygenic hypertriglyceridemias. The highest bile acid production may occur in familial type IV hyperlipoproteinemia. Cholestyramine causes a clearcut increase in fecal bile acids but to a lesser extent than in normal subjects (Table II).[5] The increase of the two primary bile acids was quite equal. Since in addition, cholesterol elimination as neutral sterols is slightly reduced the overall synthesis of cholesterol remains subnormal.

Kinetic analysis of bile acids have shown a consistently high bile acid production rate, the synthesis of cholic acid being especially elevated. Cholestyramine increased bile acid production only by about 50% almost exclusively as chenodeoxycholic acid.[9]

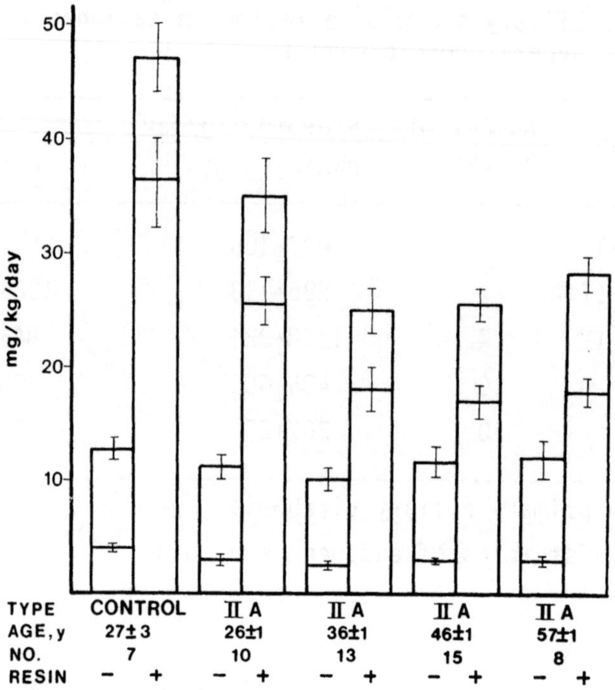

Figure 1. Fecal bile acid (lower part of columns) and cholesterol synthesis (whole column; difference between the sum of fecal bile acids plus neutral sterols and dietary cholesterol; dietary cholesterol 1.5 mg/kg/day) in normolipidemic control subjects and patients with familial hypercholesterolemia (II A) before and after cholestyramine (resin)

Secondary hyperlipidemias

Hypothyroid subjects[11] usually have a normal bile acid output, while cholesterol elimination as neutral steroids is markedly impaired. Tyroxine treatment increased elimination as neutral sterols but has no consistent effect on bile acids. Cholestyramine given to hypothyroid subjects reduces serum cholesterol and increases bile acid output comparable to that found after thyroxine treatment or in hyperthyroid patients.

In primary biliary cirrhosis biliary secretion of both cholesterol and bile acids is markedly reduced (Table III). The aborption percentage of bile acids is normal that of cholesterol low and the fecal outputs are clearly subnormal. Thus, the absolute absorption is subnormal, yet the overall cholesterol synthesis (sum of fecal bile acids plus endogenous neutral sterols) is markedly decreased.

REFERENCES

1. Sodhi H.S., Kudchodkar B.J. & Mason D.T. Adv. Lipid. Res. 1980: 17: 107.
2. Miettinen T.A., Pelkonen R., Nikkila E.A. & Heinonen O. Acta Med. Scand. 1967: 182: 645.
3. Miettinen T.A. 1973. In Bile acids. Eds. P.P. Nair & D. Kritchevsky. Plenum Press Vol. 2: 191.
4. Miettinen T.A. Arch. Surgery 1978: 113: 45.
5. Miettinen T.A. 1975. In Handbook of experimental pharmacology. Pharmacology of hypolipidemic agents. Ed. D. Kritchevsky, Springer-verlag, Berlin-Heidelberg-New York, p. 109.
6. Moutafis C.D. & Myant N.B. Clin. Sci. 1969: 37: 443; Moutafis C.D., Simons L.A., Myant N.B., Adams P.W.R. & Wynn V. Atherosclerosis 1977: 26: 329.
7. Grundy S.M., Ahrens E.H. Jr. & Salen G. J. Lab. Clin. Med. 1971: 78: 94.
8. Carter G.A., Connor W.E., Bhattacharyya A.K. & Lin D.S. J. Lipid. Res. 1979: 20: 66.
9. Hellstrom K. & Einarsson K. Clinics in Gastroenterology 1977: 6: 103.
10. Miettinen T.A. Circulation 1971: 44: 842.
11. Miettinen T.A. 1972. In Bile acids in human diseases. Ed. P. Back & W. Gerok, F.K. Schattauer Verlag, Stuttgart-New York, p. 117.

DIETARY REGULATION OF RAT SERUM LIPIDS AND LIPOPROTEINS

M. Boll, E. Bruckner, and J. Berndt

Abt. Zellchemie, Ges. F. Strahlen - u. Umweltforschung

München FRG

SUMMARY

Effects of dietary manipulation on serum lipids and lipoproteins in rats of different age are reported:

Serum lipids (triglycerides, cholesterol and phospholipids) decreased on starvation. The decrease was more intense in young than in old individuals. In contrast, serum proteins decreased only in old animals.

Refeeding a balanced or a high-sucrose, fat-free diet increased serum lipids. Triglyceride changes followed oscillatory kinetics, while cholesterol and phospholipids exhibited an overshoot phenomenon. Cholesterol and phospholipids, after starvation responded to refeeding only with a 24h retardation, instead they continued to decline after initiation of refeeding. Refeeding a balanced diet produced marked differences of the serum lipid changes between young and old animals, while the changes were very similar in the two series of rats on refeeding a fat-free diet.

Serum lipid variations on starvation and refeeding were reflected in the lipoprotein profiles. Triglyceride changes in the chylomicron-and VLDL fractions followed precisely the oscillatory kinetics, cholesterol and phospholipid variations in the LDL and HDL fractions were exactly those of total serum. In addition, VLDL cholesterol and phospholipid strongly decreased during starvation and increased immediately after initiation of the refeeding period.

INTRODUCTION

It is well established that the extent of metabolic variations observed among lipids is a function of the nutritional state and also of developmental and hormonal changes of the animal. However, the control mechanisms which modulate these changes have not yet been precisely evaluated. Two types of mechanisms are currently discussed: short term effects related to substrate and cofactor fluctuations or to activation-inactivation processes of preformed enzymes and long term (adaptive) effects involving changes in the content of enzymes.

It has recently been reported that sex and age related differences exist in the lipid levels of rat serum (1). These variations were shown to be reflected in the profiles of the serum lipoproteins (2). The present contribution describes effects of dietary manipulations on serum lipids and lipoproteins in rats with emphasis on the age of the animal.

Methods

Female Wistar rats, age 30d and 150d, were used and maintained on a standard commercial diet. For refeeding experiments, rats were fasted for 72h and subsequently refed with either the above diet or with a high-sucrose, fat-free diet (ICN Nutritional Biochemicals, Cleveland, Ohio, 901683) for different periods of time. Blood was obtained from ether anesthetized rats by abdominal aortic puncture. Lipoproteins were prepared from pooled serum of 8-12 animals by ultracentrifugation (Beckman L5-50, rotor SW 50.1) according to (3). Recovery of lipids in the fractions as compared with the serum values was 80-95 percent. Lipids were analyzed: Triglycerides (4) (Sigma, St. Louis, Kit 320 UV), cholesterol (5)(Boehringer, Mannheim, CHOD-PAP method, Kit 171 626), phospholipids (6) (Boehringer, Mannheim, kit 124974). Proteins were determined according to (7).

RESULTS AND CONCLUSIONS

Starvation decreased the levels of the serum lipids. As can be seen from Table I, the decrease was more intense in 30d than in 150d old animals. The difference between young and old rats was greatest with serum cholesterol followed by serum phospholipids and by triglycerides. In contrast, serum protein levels did not change in 30d old rats during fasting but declined in 150d old animals.

Refeeding starved animals a balanced diet restores the original status of the liver as reflected e.g. in the levels of the serum lipids which return to normal. Refeeding of high-sucrose, fat free diets is known to drastically increase the levels of biosynthetic enzymes (e.g. for triglyceride and for phospholipid synthesis) in the liver which can now rise significantly above normal values. This is accompanied by strong and rapid alterations in the fatty acid composition of the total lipid fraction and the triglycerides in the liver. In the present investigation 30d and 150d old rats were fasted for 72h and were subsequently refed with either a balanced, normal diet or with a fat-free diet for different periods of time. Chronological changes of the serum lipids were measured.

In Fig. 1 the triglyceride levels are depicted. It can be seen (part B), that refeeding with the fat-free diet caused drastic

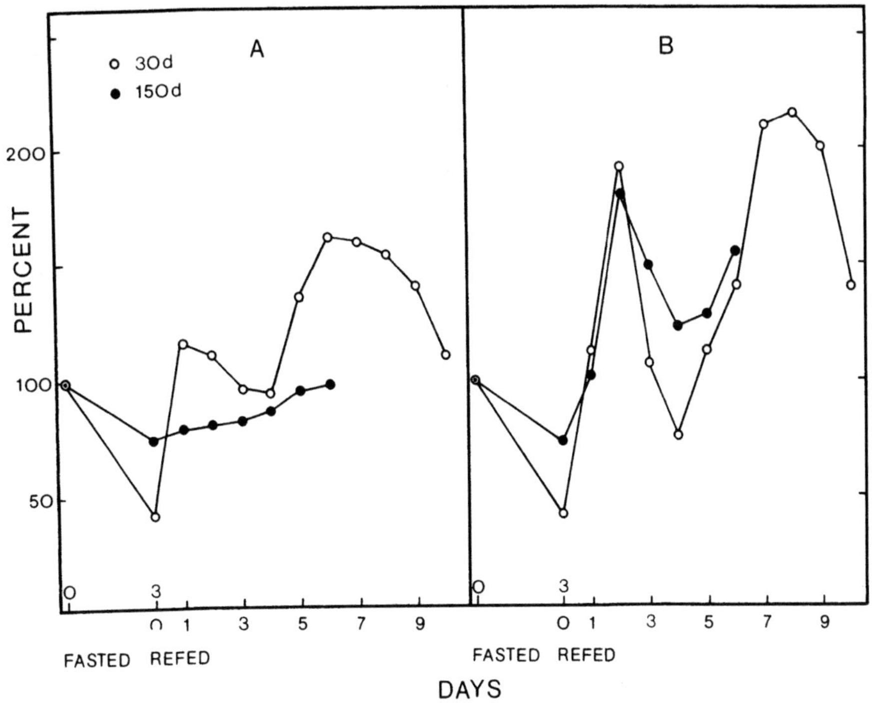

Figure 1. Changes of Serum Triglycerides in 30d and in 150d old Rats on Fasting and Refeeding a balanced, normal (A) or a fat-free (B) Diet.

The results are presented in percent with the values of normal fed control animals set at 100. Values here were 126/mg/100 ml for 30d and 230mg/100ml for 150d old animals. Each point represents average value of groups of 8-12 identically treated rats of different nutritional states. For details see "Method" section.

changes which showed oscillatory-like kinetics. These variations were similar in both young and old individuals. It is noteworthy, that the changes persisted for considerable periods of time; after 10 days of refeeding the levels had not yet again reached a steady niveau. In contrast, refeeding the rats with the balanced, normal diet (part A) resulted in a less pronounced oscillatory pattern in young individuals while old animals slowly restored their original triglyceride levels.

In Fig. 2 the same nutritional situation of fasting and subsequent refeeding is presented for the serum cholesterol. Again the changes were largely identical in both series of rats when refeeding was performed with the fat-free diet (part B), but age related differences were seen with balanced, normal refed animals (part A). In fat-free refed animals cholesterol increased sharply to levels considerably above the control. Such overshoot kinetics were also

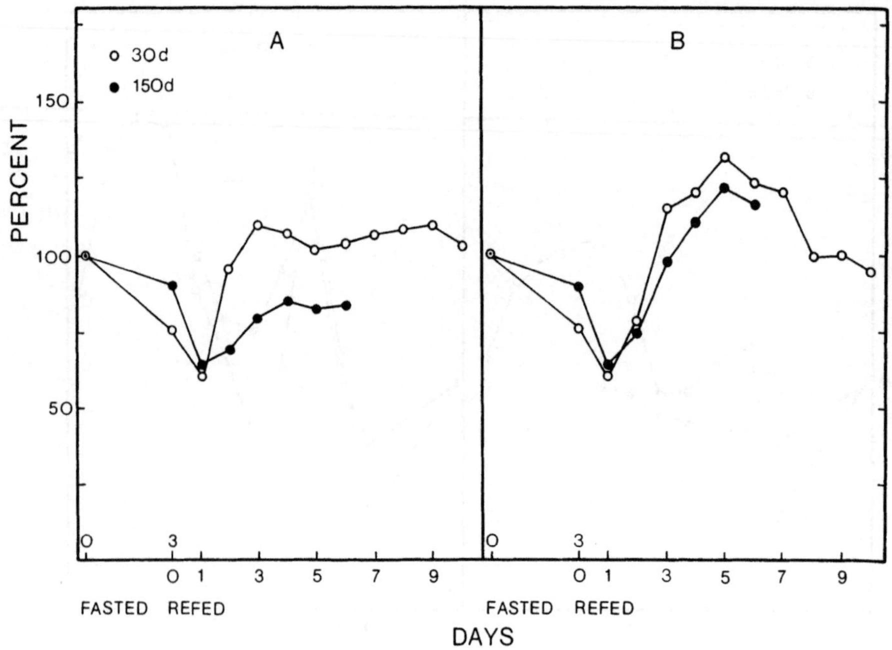

Figure 2. Changes of Serum Cholesterol in 30d and in 150d old Rats on Fasting and Refeeding a balanced, normal (A) or a fat--free (B) Diet.

For details see legend to Fig.1. Values of normal fed con-controls were 65mg/100ml for 30d and 98mg/100ml for 150d old animals.

found with the phospholipids. Here they were even more pronounced (see Fig.3). Of special interest was the observation that after the initiation of the refeeding period, cholesterol levels first continued to decrease. This further decline was found independent of the duration of the preceding starvation period. As it can also be seen from Fig. 2, the additional decrease was more intense than that which occured during the starvation and moreover old animals were more sensitive than young ones.

Fig. 3 is a summary of the variations of serum phospholipids on starvation and refeeding. The changes here were largely identical to those observed with cholesterol.

The variations of the serum lipids just described were reflected in the profiles of the lipoproteins. When comparing the triglyceride levels in the different lipoprotein fractions of normally fed, fasted and refed animals, it was seen that the levels were substantially lowered during fasting in the chylomicron-and VLDL fractions and in turn were drastically increased in these fractions after refeeding. This is shown in Fig.4. Extending the refeeding

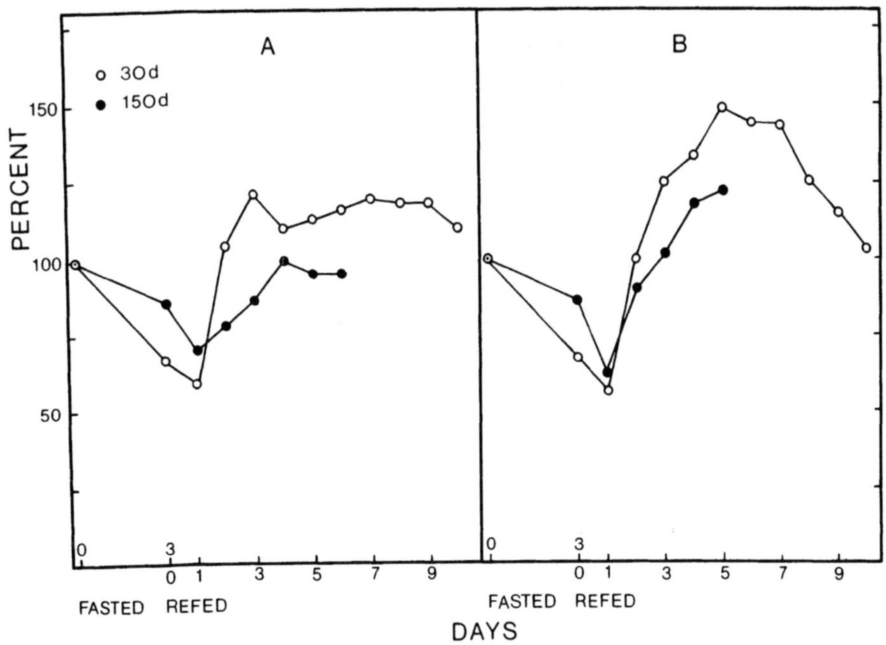

Figure 3. Changes of Serum Phospholipids in 30d and in 150d old Rats on Fasting and Refeeding a balanced, normal (A) or a fat-free (B) Diet.

For details see legend to Fig.1. Values of normal fed controls were 160mg/100ml for 30d and 268mg/100ml for 150d old animals.

period beyond 48h again showed a strong decline in these two lipoprotein fractions. Thus, triglyceride changes in the lipoprotein profiles followed precisely the oscillatory kinetics of this lipid class in the serum (see Fig.1). LDL and HDL triglycerides were unaffected by fasting and refeeding (Fig.4).

The variations of cholesterol in the lipoprotein fractions are shown in Fig.5. VLDL cholesterol, although only a small portion compared with the two mayor cholesterol-carrying fractions LDL and HDL, showed drastic changes. There was a strong decrease (80 percent) of this fraction on starvation and a likewise marked increase after the onset of refeeding. On the other hand, cholesterol in the LDL and HDL fractions declined only slightly during fasting and continued to decrease when refeeding was initiated. Only 48h after refeeding started, LDL and HDL cholesterol increased again. Thus, these two fractions represented the changes seen with serum cholesterol.

Variations of the phospholipids in the lipoprotein fractions on starvation and refeeding were similar to those of cholesterol. This is shown in Fig.6.

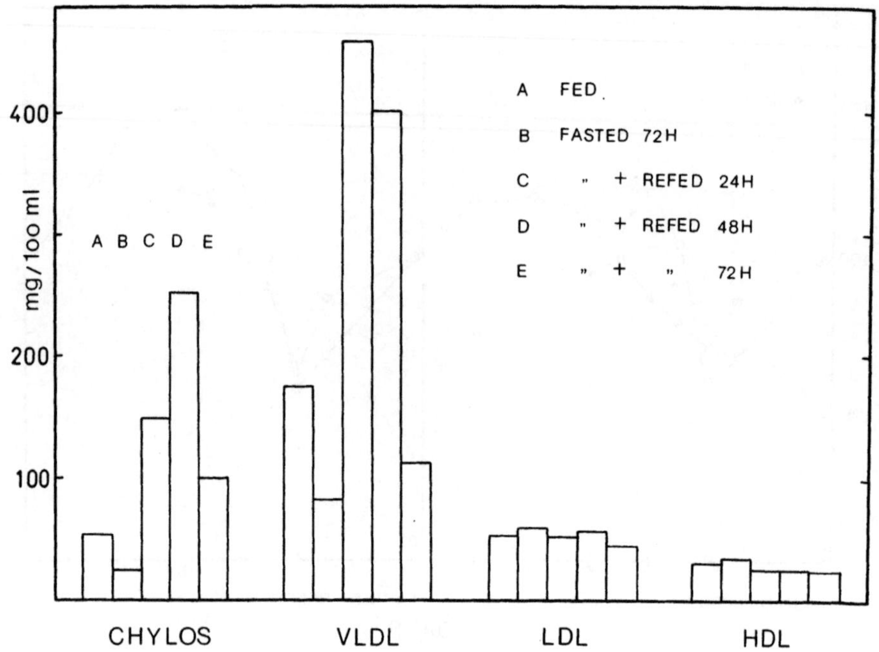

Figure 4. The Effect of Starvation and Refeeding on the Triglyceride Composition of the Lipoprotein Fractions.

Lipoproteins were prepared from serum of groups of 8-12 rats, age 30d, of different nutritional states and lipids were analyzed in the fractions as described. Refeeding was with the fat-free diet. For details see "Method" section.

In conclusion, dietary manipulations can markedly affect serum lipid levels. It is of interest that the biosynthesis of triglycerides and of cholesterol and phospholipids and their appearance in the serum were considerably different in response to the dietary state but that cholesterol and phospholipids always behaved identical. Lipid variations persisted for unexpectedly long periods of time. Changes of serum lipids were also found to be a function of age, thus, developmental differences exist in the metabolism of hepatic lipids. Old individuals but not young ones have different capacities for restoring their normal lipid levels in the serum under the variuos refeeding conditions. All these observations indicate that different types of control mechanisms will be necessary for the precise adjustment of the various lipid levels.

Dietary variations have an obvious counterpart in the changes of the lipoprotein levels and their composition. Moreover, modifications of specific lipoprotein fractions can be identified.

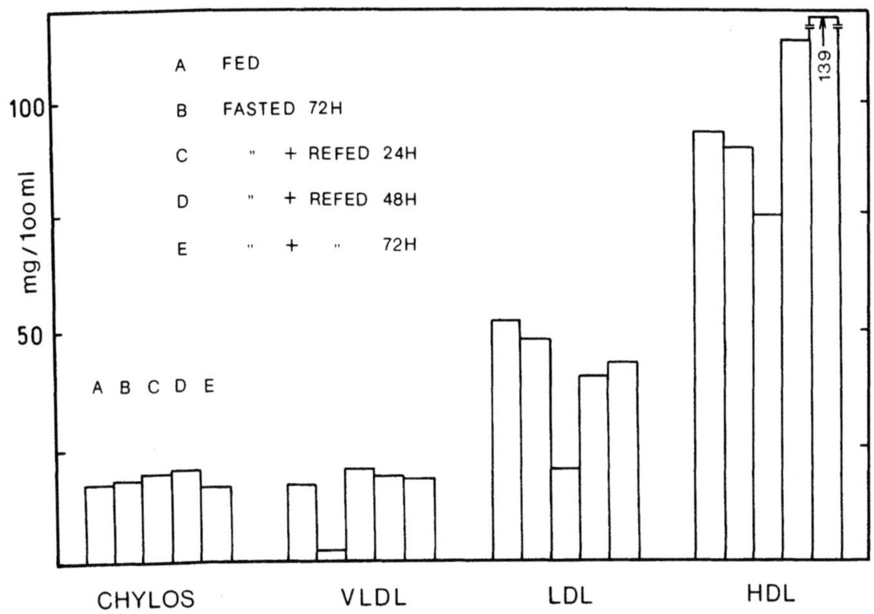

Figure 5. The Effect of Starvation and Refeeding on the Cholesterol Composition of the the Lipoprotein Fractions. For details see legend to fig. 4.

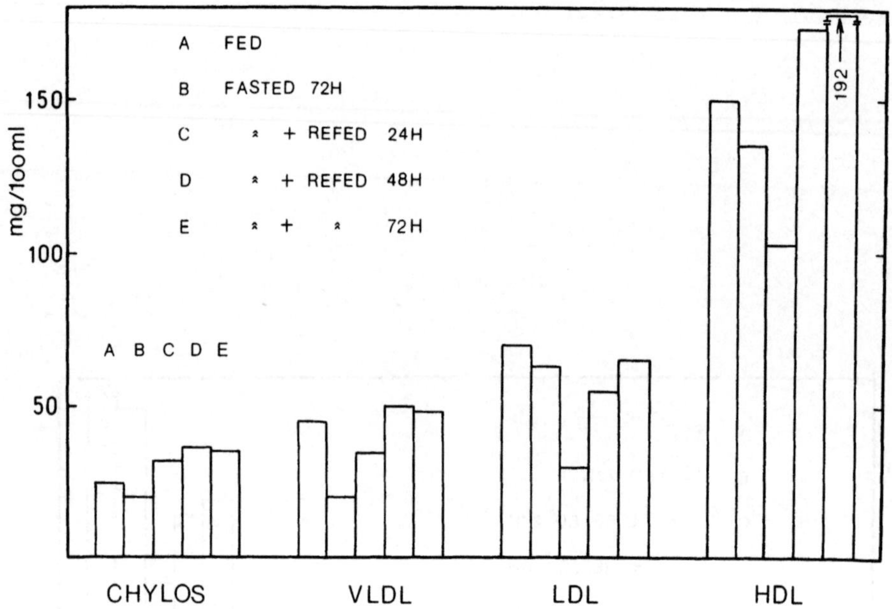

Figure 6. The Effect of Starvation and Refeeding on the Phospholipid Composition of the Lipoprotein Fractions.

For details see legend to Fig. 4.

Table I. Decrease of serum lipids and protein in young (30d) and in old (150d) rats after 72h of starvation

	Percent Decrease	
	30d	150d
TRIGLYCERIDES	56	40
CHOLESTEROL	28	10
PHOSPHOLIPIDS	32	18
PROTEIN	<5	15

REFERENCES

1. Boll, M., Bruckner, E. and J. Berndt, (1980) Hoppe-Seyler Z.f. Physiol. Chem. 361,223
2. Boll, M., Bruckner, E. and J. Berndt, (1980) 13th FEBS Meeting Jerusalem, Book of Abstracts p.210,C6-P6
3. Hatch, F.T. and R.S. Lees, (1968) in "Adv. in Lipid Res." " (Paoletti, R. and D. Kritchevsky,eds.) Vol. 6,pp. 1-68, Academic Press, New York and London
4. Bucolo, G. and H. David (1973) Clin. Chem. 19,476
5. Rochlau, P., Berndt, E. and W. Gruber (1974) J.Clin.Chem. Clin. Biochem. 12,403
6. Zilversmit, D.B. and A.K. Davis (1950) J.Lab.Clin.Med. 35,155
7. Lowry, O.H., Rosebrough, N.J., Farr, A.L. and R.J. Randall (1951) J.Biol.Chem. 193,265

MULTIVARIATE ANALYSIS OF SERUM APOLIPOPROTEIN LEVELS AND OTHER RISK FACTORS FOR MYOCARDIAL INFARCTIONS

G. Fager, O. Wiklund, S.-O. Olofsson, L. Wilhelmsen, and G. Bondjers

Department of Medicine I, Sahlgren's Hospital
S-413 45
Göteborg, Sweden

INTRODUCTION

Recently, much evidence has accumulated that apolipoproteins have specific functions in the intermediary metabolism of lipids (for review and references see; Fager, 1979). Therefore, we wanted to test the hypothesis that serum apolipoproteins might be more specific and sensitive markers of the dyslipoproteinemias involved in the early development of myocardial infarction. This paper reports the results of multivariate analyses of data on serum apolipoproteins A-I (apoA-I), A-II, B (apoB) and D (apoD) as well as on serum lipids, alphalipoprotein cholesterol and several other variables discussed in relation to the development of myocardial infarction.

SUBJECTS AND METHODS

The myocardial infarction group consisted of 25 survivors of a first myocardial infarction occuring between 1975 and 1977 and when the subjects were 40-44 years of age. They comprise 71 per cent of all first myocardial infarctions and 86 per cent of all survivors within the particular subpopulation. A control group was selected at random from the municipal census list of the corresponding subpopulation. The 76 participants in this group represented 74 per cent of the selected men.

Smoking habits and the levels of physical activity at work and during leisure time were evaluated from standard questionnaires. Information about men registered for alcohol intemperance was obtained from the local Temperance Board. Blood pressure was measured after infarction in the survivors. Therefore, blood pressure was omitted in the statistical calculations. Body weight index was calculated as; body weight (body height-100) kg/cm.

Blood samples were drawn in the morning after 12 hours of non-smoking and fasting and after 48 alcohol-free hours. Serum cholesterol (Borner and Klose, 1977) and triglyceride (Wahlefeld, 1974) levels were determined by fully enzymatic procedures. Alphalipoprotein cholesterol was measured according to Leppänen (1956) on the clear supernate after precipitation of serum with heparin and manganese chloride (Wilkund et al. 1980 a). Serum levels of apoA-I, A-II, apoB and apoD were determined by electro-immunoassays as reported in detail elsewhere (Fager, 1979, Fager et al. 1979, 1980 a and b, Wiklund et al. 1980 b).

Differences between groups were tested with the Mann-Whitney U or Chi-square tests (Armitage, 1974). Correlations were evaluated as Spearman's rank correlation coefficients (Siegel, 1956). Multiple regression and discriminant function analyses were carried out according to Snedecor and Cochran (1972). The relative importance of different variables at multiple regression was evaluated from the standard partial regression coefficient (standard b=partial regression coefficient/its standard deviation).

RESULTS

Serum apoA-I (p=0.003) and A-II (p=0.0003) levels as well as alphalipoprotein cholesterol levels (p=0.0002) were all significantly lower in infartion survivors than in controls as evaluated by simple statistics. These variables were strongly intercorrelated and at multiple regression analyses only the A-II levels were independently correlated to infarction (standard b=-3.12).

Serum triglyceride (p=0.002) and apoB (p=0.012) levels were higher in survivors by simple statistics, whereas only serum triglycerides were a variable of independent importance for infarction in the multiple regression analyses (standard b=2.06). Serum cholesterol and apoD levels discriminated infarction survivors by neither statistical procedure.

A high tobacco consumption before infarction characterized infarction survivors by simple statistics (p=0.0001) as well as by multiple regression analyses (standard b= 2.64).

The levels of physical activity during leisure time were of borderline significance at multiple regression analyses (standard b=1.83), although it did not fall out significantly at simple statistical analyses. Registration at the Temperance Board was more common among survivors than among controls as evaluated by the chi-square test (p=0.003) and it was of borderline significance at multiple regression analyses (standard b= 1.83).

The multiple regression coefficient (R=0.53, p=0.00001) explained 28 per cent of the variation of the dependent variable when only significant independent variables (serum A-II and triglycerid levels and tobacco consumption) were included. By adding the borderline variables (physical activity during leisure time and registration at the Temperance Board) the multiple R (0.57, p=0.00001) did not increase significantly.

The significant variables from multiple regression analyses were entered into a discriminant function. The results gave a good separation between the group of lowest and the group of highest relative prevalence of infarction survivors. The curve rose continously from zero to more than six times the average prevalence with increasing values of the function. Fifty per cent of all infarction survivors corresponded to the upper decile class of the controls. The inclusion of variables of borderline significance at multiple regression analyses did not improve the separation between cases and controls by the discriminant function.

A predictive hypothesis may be derived from the present results and subjected to testing in future prospective studies. One may assum that the lipoprotein data are representative also for candidates of future infarction. The assumption may then be tested along with the discriminant function of the present study. The expectation would be that the average incidence of first infarction (102.3 per 100,000 males and year in the present study) would increase from zero to more than 600 per 100,000 males and year with increasing values of the function. Furthermore, the upper decile class would contain 50 per cent of the infarctions to come.

DISCUSSION

The present study gives the first report on multivariate analyses of several apolipoproteins and other lipoprotein variables and risk factors for myocardial infarction in a well-defined population-based case control study.It supports the hypothesis that apolipoproteins are more specific and sensitive markers of the dyslipoproteinemias involved in the early development of myocardial infacrtion than are the levels of serum lipids. Serum A-II levels were not only more important in discriminating survivors of infarction from controls than other lipoprotein variables of the study. They were also more important than the established major risk factor of tobacco smoking. Furthermore, the A-II levels were almost as good in discriminating cases and controls as was a discriminant function including all significantly discriminative variables of the study (serum A-II and triglyceride levels and tobacco consumption). Although this discriminant function gave a good separation between cases and controls, the corresponding multiple regression function explained only 28 per cent of the variation of the infarction/no infarction variable.

Due to the lack of pre-infarction registrations, blood pressure levels were omitted from the multivariate analyses. The lack of difference in blood pressure levels between survivors and controls does not contradict blood pressure as a major risk factor for infarction (Gordon et al. 1971. Wihelmsen et al. 1973), since a reduction in blood pressure has been observed in relation to infarction (McCall et al. 1979).

However, the inclusion of blood pressure in a prospective study is necessary and will probably improve the discrimination between cases and controls as well as the degree of explanation.

In conclusion, serum A-II levels may well be a risk factor for the early development of myocardial infarction even more important than the established risk factors to tobacco smoking and hyperlipidemia.

ACKNOWLEDGMENTS

The present study was supported by grants from the Medical Research Council (project No. 4531), the Bank of Sweden Tercentenary Fund and the Medical Society of Göteborg

REFERENCES

1. Armitage, P.-Statistical methods in medical research. Third printing. Blackwell Scientific Publication, Oxford, 1974.
2. Borner, K. and Klose, S.-J. Clin Chem. Clin. Biochem. 15, 121, 1977.
3. Fager, G. - Serum apolipoprotein levels in relation to acute myocardial infarction and its established risk factors. Diss., 1979, University of Göteborg, Göteborg, Sweden. Printed by GOTAB, Kungälv, Sweden.
4. Fager, G., Wiklund, O., Olofsson, S.-O., Norfeldt, P., Wilhelmsen, L. and Bondjers, G.-Artery 6 (3), 187, 1979.
5. Fager, G., Wiklund, O., Olofson, S.O., Norfeldt, P-L., Vedin, A. and Bondjers, G. -Scand. J. Clin. Lab. Invest. 40:451, 1980 a.
6. Fager. G., Wiklund, O., Olofsson, S.O., Wilhelmsson, C. and Bondjers, G. - Atherosclerosis 36:67, 1980 b.
7. Gordon, T., Sorlie,P. and Kannel, W.B.- The Framingham Study. An epidemiological investigation of cardiovascular disease. Section 27. US department of health, education and wellfare. 1971.
8. Leppänen, V. -Scand. J. Clin. Lab. Invest. 8:201, 1956.
9. McCall, M., Elmfeldt, D., Vedin, A., Wilhelmsson, C., Wedel, H. and Wilhelmsen, L.- Acts Med. Scand.
10. Siegel, S.-Nonparamentric statistics for behavioural sciences. McGraw-Hill, New York, 1956.
11. Snedecor, G.W. and Cochran, W.G. -Statistical Methods, Sixth Edition. The Iowa State University Press, Ames, Iowa, 1972.
12. Wahlefeld, A.W. - in Bermayer, HU (ed.), Methods of Enzimatic Analysis. 2end Edition. Academie Press Inc., New York and London, 1974.
13. Wiklund, O., Fager, G., Craig. I.H. Wilhelmsen, L.- Scand. J. Clin. Lab. Invest. 40.239, 1980 b.

14. Wilhelmsen, L., Wedel, H. and Tibblin, G. - Circulation 48:950, 1973.

CONTRIBUTION TO THE STUDY OF THE LIPID CONSTITUTION IN LIVER AND BRAIN HOMOGENATES OF MICE GENETICALLY OBESE

Armando Sena*, Gérard Rebel**, Robert Bieth, and
Albert Waksman***

Dept. of Bioche. of Fac. of Med. Sciences and Center
for Lipid Research from I.N.I.C.
Campo de Santana 130 - 1100 Lisbon - Portugal

SUMMARY

Lipid composition was compared in liver and brain from genetically obese (ob/ob), heterozygote (ob/+) and normal (+/+) mice. In liver, but not in brain, a constitution favoring a "rigidification" of the membranes was found in (ob/ob) mice. In both organs a decrease of the total ganglioside sfalic acid concentration was observed. These results could be related to the reduced number of insulin receptors in (ob/ob) mice hepatocytes and to a different behaviour of their brain receptors. In addition, they suggest that an abnormal complex carbohydrate metabolism involving gangliosides may exist in this syndrome.

INTRODUCTION

Genetically obese mice (ob/ob) from C57 BL/6J strain display marked overeating, hyperinsulinemia and transient or late-onset hyperglycemia. This syndrome is caused by a single recessive gene (ob), located on chromosome 6; heterozygote mice (ob/+) are clinically indistiguishable from normal mice (+/+) (1).

Modulation of membranes fluidity by insulin could be an important regulatory factor (2, 3). In its physiological action. Kahn et al. found a decreased number of insulin receptors in plasma membrane hepatocytes of(ob/ob) mice (4). More recently, Havrankova et al. (5)

suggested a different behaviour of liver and brain receptors for insulin in same mice. As lipids are know to be important in the regulation of membrane dynamicas and fluidity (6), the different behaviour of brain and liver receptors could be related to a difference in the lipid composition. Therefore, we have undertaken a study of brain and liver cholesterol to phospholipid molar ratio and phosphatide fatty acid composition, known to be the two most important factors in the regulation of lipid fluidity. Ganglioside composition was also considered owing to the importance of these compounds in membrane recognition processes and the evidence for an anomaly of complex glycoconjugates in insulin resistence (7).

MATERIAL AND METHODS

The three genetic lines of C57 BL/6J mice came from the "Centre d'Elevage du CNRS" (45 - Orléans-la-Source-France). They were maintained under the same conditions of light and temperature and fed ad libitum. Mice aged between 5 and 7 months were killed by decapitation and brain and liver were removed and immerged immediately in liquid nitrogen. Organs were kept at - 20° C, lyophilized at least during 12 hours and dry weight determined.

Lipids were extracted and partitioned according Folch et al. (8). Lower phase of Folch was washed twice with chloroform/methanol/water (3:48:5, v/v/v) and these two phases were added to the upper phase of Folch. This process permits to recuparate pratically all the gangliosides that remained in the lower phase. The lower phase was evaporated in vacuum and neutral lipids and phospholipids separated according Hirsch and Ahrens (9). Cholesterol was assayed with Idler and Baumann technique (10) and lipid phosphorus was measured according to Macheboeuf and Delsal (11). Phospholipid fatty acids were obtained after alkaline methanolysis of the lipid extract (12) and they were separated by gas-liquid chromatography on a column of 10% EGS-Sx on chromosorb WHP. The identification of the peak was performed as described in (13). Gangliosides were purified by the method of Wells and Dittmer (14) as modified in our laboratory and were assayed with the technique of Miettinen and Takkiluukainem (15), separation was realized by thin layer chromatography using three solvents according Harth et al. (16). The distribution of sialic acid among gangliosides was obtained by densitometry according to Smid and Reinisova (17) after their visualization on the plate with Biol's orcinol reagent

RESULTS AND DISCUSSION

Livers from (ob/ob) mice have at least the double of the weight of that of their lean littermates (ob/+ and +/+). Brains from (ob/ob) and (ob/+) have a similar weight which is around 35% less than that of the genetically normal animals. There was no sexual differences neither in the weight nor in the chemical composition of the studied organs.

In liver, cholesterol and phospholipid concentrations are very similar in (+/+) and (ob/+) mice whereas in obese animals cholesterol is higher by 40% and the corresponding cholesterol/phospholipid molar ratio increases by 77% (Table Ia). In brain, however, phospholipid decreases only by 30% in both lines of mutant mice whereas cholesterol concentration falls by 40%.Accordingly, there is only a slight decrease of cholesterol to phospholipid molar ratio in brain (Table Ib). This ratio is assumed to be an important factor in the regulation of the fluid state of membranes. Its notable elevation in(ob/ob) mice liver could thus lead to an increased"rigidification of their membranes. The consequence of the expression of (ob) gene for the cholesterol/phospholipid molar ratio in plasma membranes could be very different in liver and brain of these mice. Whereas in the normal (+/+) mice this ratio is 45% higher in brain, in (ob/ob) mice this value is inversed and is 43% higher in liver. In brain, a net reduction of total cholesterol and phospholipids may be an important contribution to its reduced dry weight in(ob/+) and (ob/ob) mice. Tables IIa and IIb present relative percentages of phosphatide fatty acids which were found to be modified. The ratio between the polyunsatured and the monounsaturated fatty acids is strongly decreased only in liver (Table IIa). The decrease of this value (60%) could contribute to the elevated membrane microviscosity in (ob/ob) mice liver but is already clearly present in heterozygote (ob/+) mice (35%). The most important difference is a 49% increase of 18:1 (oleic acid) in (ob/+) and a much larger increase of 106% in (ob/ob) mice. Linoleic acid (18:2) is reduced by 13% in the heterozygote and by 35% in obese mice. As far as gangliosides are concerned a significant reduction of total sialic acid concentration was observed in liver (39%) (Table IIIa) and in brain (38%) (Table IIIb) of (ob/ob) mice. However, the presence of one mutant gene (ob/+) already effects these complex glycolipids mainly in brain (-18%). Although we could not find any difference in the ganglioside pattern between normal and mutant lines, GM2 represents around 80% of the total gangliosides in liver of these mice. Nevertheless, ganglioside composition in brain is quite typical of other mammalians.

Massa et al. (2) and Moreno and Farias (3) conclude that insulin may act by reducing membrane fluidity. Several studies (19, 20,21) proposed that the physical state of membranes may modulate the exposure of their hormonal receptors. As a matter of fact Luly and Shinitzky (22) showed that an enrichment of plasma membranes with cholesterol avoids the change in lipid microviscosity normaly induced by insulin and decreases the number of available receptors for the hormone. It is tempting therefore to speculate that the abnormal lipid constitution of (ob/ob) hepatocyte plasma membranes suggested by our results could account at least in part for their reduced insulin receptor number (4). Havrankova et al. (5) have found that the brain insulin receptor content of these mice was indistinguishable from that of their littermates, whereas the receptors in liver cells were reduced by 40-50%. Although an independent regulation of brain and liver receptors, as proposed by these authors, cannot be excluded, the different lipid composition of the two organs suggests, however, that the (ob) gene might express itself very differently according to the implied target cell. This conclusion is also supported by the observations of Hyslop and York (23) about the lipid constitution of (ob/ob) adypocite plasma membranes. Some contradictory data also exist on insulin influence upon lipid viscosity in liver and adypocite membranes (22). The specific physico-chemical behaviour of the microenvironment in membranes of different tissues could provide another possible explanation for the observed differences (24) and could account for the peculiar reactions of different membranes towards a same physiological ligand. In a recent report (25) evidence was forwarded for a possible different organization of liver and adipocyte insulin receptor. While this paper was in preparation Rouer et al. (26) presented data on an increase of 18:1 fatty acid and a decrease of 18:2 fatty acid in phospholipids of (ob/ob) liver microsomes. Our results confirm this data and demonstrate that the abnormal fatty acid composition really exists even in heterozygote (ob/+) mice, suggesting that it could be linked to the development of insulin resistance. As a matter of fact, one major conclusion of our findings is that although clinicaly normal and insulin responsive, (ob/+) mice have already several biochemical alterations and should not be used as experimental controls when testing (ob/ob) mice. Both studies support the view that unsaturated fatty acids may be implicated in insulin transmembrane communication (27) and in this way it is relevant that an alteration of fatty acid side chains of membrane phospholipids can alter the expression of glycoprotein receptors (28).

Finally, results concerning gangliosides may reflect a more general dysfunction of intracellular complex carbohydrate metabolism and glycosidation systems. Supporting this view, is the fact that insulin shocks and seizures can reduce neutral glycolipids in brain (29). In addition, an abnormal glycoprotein metabolism in genetically obese mice has been proposed (7) as well as changes in glycosyltransferases in diabetic patients (30). Jakobsen and Sidenius (31) could show a reduced axonal transport of glycoproteins in brain of early experimental diabetes. These findings support the hypothesis that an early general abnormal complex carbohydrate metabolism involving not only glycoproteins but also glycolipids such as gangliosides could be implicated in the development of these syndromes.

ACKNOWLEDGMENT

A. Sena is grateful to "Fundação Calouste Gulbenkian". The authors are greateful to Mrs Paulette Guérin and Marie-Claude Schricke for their technical assistance and to Mrs. Claudine Thomassin for careful taping this manuscript.

Part of this work was supported by the "Institut National de la Santé et de la Recherche Médicale", grant nº 803022 and by the "Fondation pour la Recherche Médicale Française".

Table Ia. Cholesterol and phospholipid in liver of (+/+), (ob/+) and (ob/ob) mice

	+/+ (2)	ob/+ (3)	ob/ob (3)
Cholesterol*	11.78 ± 2.31	10.15 ± 3.01	16.49 ± 2.73
Phospholipid **	2.01 ± 0.25	1.87 ± 0.24	1.62 ± 0.42
Molar ratio			
Cholesterol / phospholipid	0.47	0.43	0.83

* Values are expressed in μgrams per mg of dry weight and represent the mean ± SEM of (2) to (3) determinations.

** Values are expressed in μgrams lipid phosphorus per mg of dry weight.

Table Ib. Cholesterol and phospholipid in brain of (+/+), (ob/+) and (ob/ob) mice

	+/+ (2)	ob/+ (3)	ob/ob (3)
Cholesterol*	49.48 ± 1.07	30.13 ± 1.08	29.25 ± 1.15
Phospholipid**	5.84 ± 0.08	4.11 ± 0.46	4.01 ± 1.30
Molar ratio			
cholesterol / phospholipid	0.68	0.58	0.58

* Values are expressed in μgrams per mg of dry weight and represent the mean ± SEM of (2) to (3) determinations.

** The same as in Table Ia.

Table IIa. Phosphatide fatty acid composition in liver of (+/+), (ob/+) and (ob/ob) mice

	+/+ (2)	ob/+ (3)	ob/ob (3)
16:1*	3.10±1.70	2.90±1.22	4.37±0.87
18:1	10.20±0.99	15.20±1.01	21.03±3.47
18:0	16.85±1.06	16.30±1.28	14.27±2.15
18:2	17.85±3.04	15.60±3.04	11.53±2.20
20:4ω6	14.30±2.59	12.63±1.70	11.20±1.63
22:6ω3	12.55±4.50	10.97±2.94	11.53±1.31
Ratio **			
poly/mono	3.33	2.16	1.35

* Values are the means ± SEM from (2) to (3) animals and expressed as a percentage of total phospholipid fatty acids.

** Ratio between polyunsaturated and monounsaturated fatty acids.

Table IIb. Phosphatide fatty acid composition in brain of (+/+), (Ob/+) and (ob/ob) mice

	+/+ (2)	ob/+ (3)	ob/ob (3)
16:1*	1.80±0.85	1.43±0.59	2.17±0.91
18:1	23.05±0.64	23.47±1.39	22.10±3.20
18:0	19.90±0.14	21.57±1.60	19.13±0.96
18:2	1.10±0.42	0.87±0.15	1.13±0.47
20:4ω6	9.20±0.85	10.27±1.96	8.67±0.84
22:6ω3	15.30±4.95	18.50±5.77	15.07±1.56
Ratio**			
poly/mono	0.97	1.19	1.02

* Values are the means ± SEM from (2) to (3) determinations and expressed as a percentage of fatty acids found modified in liver.

** The same as in Table IIa.

Table IIIa. Ganglioside sialic acid and ganglioside pattern in liver of (+/+), (ob/+) and (ob/ob) mice

Gangliosides	+/+ (2)	ob/+ (3)	ob/ob (3)
Total sialic acid*	0.51±0.01	0.46±0.02	0.31±0.05
GM3**	17.65	17.69	18.90
GM2	81.86	76.36	81.12
GM1	1.04	1.81	1.65

* Values are the means ± SEM from (2) to (3) determinations and expressed in μgrams per mg of dry weight.

** Individual gangliosides are expressed as the percentage of total gangliosides. The nomenclature of Svennerholm was used (18).

Table IIIb. Ganglioside sialic acid and ganglioside pattern in brain of (+/+), (ob/+) and (ob/ob) mice

Gangliosides	+/+ (2)	ob/+ (3)	ob/ob (3)
Total sialic acid*	4.33±0.04	3.54±0.10	2.70±0.31
GM3**			
GM2	7.33	6.86	6.98
GM1	14.56	13.42	14.39
GD3	3.20	3.63	3.51
GD1a	25.24	25.61	25.44
GD1b	12.66	12.29	11.53
GD2	9.77	8.90	8.93
GT1	21.80	21.83	21.72
GQ1	7.45	6.85	7.54

* and ** : the same as in Table IIIa

REFERENCES

1. Coleman, D.L., and Hummel, K.P. (1973) Diabetologia 9, 287-293.
2. Massa, E.M. Morero, R.D., Bloj, B., and Farias, R.N. (1975) Biochim. Biophys. Res. Commun. 66, 155-162.
3. Moreno, H., and Farias, R.N. (1976) Biochim. Biophys. Res. Commun. 72, 74-80.
4. Kahn, C.R., Neville, Jr., D.M., and Roth, J. (1973) J. Biol. Chem. 248, 244-250.
5. Havrankova, J., Roth, J., and Brownstein, M.J. (1979) J. Clin. Invest. 64, 636-642.
6. Cremel, G., Rebel, G., Warter, J.M., Rendon, A., and Wasksmar A. (1976) Arch. Biochem. Biophys. 173, 255-263.
7. Chang, K.J., Huang, D., and Cuatrecasas, P. (1975) Biochem. Biophys. Res. Commun. 64, 566-573.
8. Folch, J., Lees, M., and Sloane-Stanley, G.H. (1957) J. Biol. Chem. 226, 497-509.
9. Hirsch, J., and Ahrens, E.H. (1958) J. Biol. Chem. 233, 311-320.
10. Idler, D.R., and Baumann, C.A. (1953) J. Biol. Chem. 203, 389-396.
11. Macheboeuf, M., and Delsal, J. (1943) Bull. Soc. Chim. Biol. 25, 116-120.
12. Dawson, R.M.C. (1960) Biochem. J. 75, 45-53.
13. Robert, J., Rebel, G., and Mandel, P. (1977) Biochimie 59, 417-423.
14. Wells, M.A., and Dittmer, J.C. (1963) Biochemistry 2, 1259-1263.
15. Miettinen, T., and Takki-Luukkainen, I.-T. (1959) Acta Chem. Scand. 13, 856-858.
16. Harth, S., Dreyfus, H., Urban, P.F., and Mandel, P. (1978) Anal. Biochem. 86, 543-551.
17. Smid, F., and Reinisova, J. (1973) J. Chromatogr. 86, 200-204.
18. Svennerholm, L. (1963) J. Neurochem. 10, 613-623.
19. Shinitzky, M. and Inbar, M. (1976) Biochem. Biophys. Acta 433, 133-149.
20. Borochov, H., and Shinitzky (1976) Proc. Natl. Acad. Sci. USA 73, 4526-4530.
21. Shinitzky, M., and Rivnay, B. (1977) Biochemistry 16, 982-986.
22. Luly, P., and Shinitzky, M. (1979) Biochemistry 18, 445-450.
23. Hyslop, P.A., and York, D.A. (1980) Biochem. Biophys. Res. Commun, 92, 829-824.
24. Sauerheber, R.D., Lewis, U.J., Esgate, J.A., and Gordon, L.M. (1980) Biochem. Biophys. Acta 597, 292-304.
25. Schweitzer, J.B., Smith, R.M., and Jarret, L. (1980) Proc. Natl. Acad. Sci. USA 77, 4692-4696.
26. Rouer, E., Dansette, P., Beaune, P., and Leroux, J.L.

(1980) Biochem. Biophys. Res. Commun. 95, 41-46.
27. Mukherjee, S.P., Mukherjee, C., Yeager, M.D., and Lynn, W.S. (1980) Biochem. Biophys. Res. Commun. 94, 682-689.
28. Lee, G., Consiglio, E., Habig, W., Dyer, S., Hardegree, C., and Kohn, L. (1978) Biochem. Biophys. Res. Commun. 83, 313-320.
29. Mesdjian, E., Rebel, G., Mesdjian, J.L., Freysz. L., and Mandel, P. (1978) Neuroch. Res. 3, 125-134.
30. Cherbit, G., and Roux, E. (1978) Cybernetica 21, 339-346.
31. Jakobsen, J., and Sidenius, P. (1979) J. Neurochem. 33, 1055-1060.

EFFECT OF DIURNAL DISTRIBUTION OF FOOD INTAKE ON 24-HOUR PROFILES OF PLASMA LIPOPROTEINS

Jurgen Schneider

Department of Internal Medicine
Phillips-University
Marburg, 355, German Federal Republic

ABSTRACT

The lipid infiltration theory of atherogenesis accepted, 24 hour lipoprotein profiles may be more relevant than preprandial morning samples. Such profiles were performed in 12 metabolically healthy volunteers during two dietetic regimes identical in total food content but differing in the distribution over the day: form A meant an evening meal of 15% of total caloric intake, form B of 40%. After one week of each form, 24 hour lipoprotein profiles differed significantly in the time course of triglyceride rich lipoproteins and in the mean values over 24 hours in VLDL and LDL phospholipids and HDL cholesterol. These findings are cautiously interpreted as possible signs of differences in the catabolism of triglyceride rich lipoproteins, remnants and intermediate lipoproteins. The difference in HDL cholesterol which was higher in form A is discussed in the context of recent epidemiologic evidence.

INTRODUCTION

With developing methodology relations between dietary advice, plasma lipids or lipoproteins and atherogenesis have become complicated - 25 years ago it seemed sufficient to lower elevated total cholesterol, now in view of recent results we have to check the influence of intervention on lipoproteins which indicate a positive or negative risk to develop permature atherosclerosis. So long new concepts on atherogenic and possibly antiatherogenic properties of dofferent lipoproteins have been discussed, no data have been published dealing with the influence of the distribution of food intake over the day on lipoprotein fractions. Furthermore reports on lipoproteins and dietetic intervention usually consists of data obtained from morning (fasting) lipoprotein values. A stable level of possibly atherogenic and antiatherogenic lipoproteins over 24 hour is - silently-taken for granted. From a pathophysiological point of view and a causal role of plasma lipoproteins in atherogenesis accepted, 24 hour profiles of HDL as well as VLDL and LDL may provide a better rationale for dietetic advice than fasting values (1).

The aim of this study was to achieve preprandial lipoprotein values and 24 hour lipoprotein profiles under a defined diet, differing exclusively in the distribution of food intake during the day.

METHODS

12 volunteers, 4 females and 8 males, mainly medical students of 22 to 36 years without clinical disease inclusive overweight of more than 5% in Broca index, eating normal food of the region quoted their total nutritient intake at home to maintain their actual body weight, which was controlled daily (table I). The food intake was estimated by dietary recall data and consisted of 125 KJ/KG body weight in the mean (table I), about 40% carbohydrate, 15% protein and 45% fat with a P/S ratio of 0.3-0.5.

The test conditions for one week each resembled to the data obtained and differed in the distribution of food intake during the day: for one week the evening meal at 5.30 p.m. consisted of 15% of total caloric intake (form A), for another week of 40% (form B).

The sequence of forms A and B was alternative, 6 persons began with form A, six with form B, with 2 or 3 weeks of isocaloric intake and free distribution interposed. All 12 persons were subjected to forms A and B. The last day of each week (forms A and B) was spent in a metabolic ward, the food performed by the clinic dieticians. Throughout the two test periods daily exercise, smoking habits and intake of not more than three alcoholic beverages was kept constant.

Blood samples were taken by venous puncture at times 4, 8, 11^{00} a.m. and 2, 5, 8, 12^{00} p.m. Meals were ingested at 8 and 11^{00} a.m., 2 and 5.30 p.m., after blood sampling. In all samples determination of total serum cholesterol (2), triglycerides (3,4) and phospholipids (5) chylomicron triglycerides, separation of VLDL, LDL and HDL by preparative ultracentrifugation (6), and measurement of cholesterol, triglycerides and phospholipids in these fractions was performed. The 24 hour mean values were corrected for the different intervals during day and night and all data obtained during form A and B compared by the Wilcoxon-test.

RESULTS

The different distribution of food intake in forms A and B resulted in a shift of the peaks in total and chylomicron triglycerides from 2 to 8^{00} p.m. The differences were significant ($p<0.01$) for total triglycerides at 2 and 8^{00} p.m. with a tendency at the same points for chylomicron triglycerides (fig. 1,2) and for VLDL triglycerides (fig. 3). The corrected mean values of both parameters

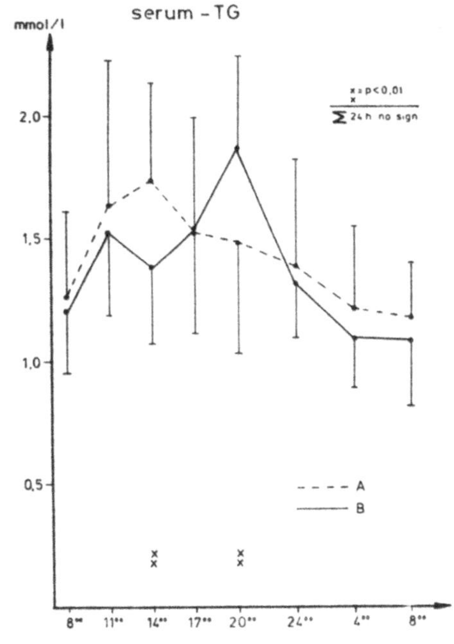

Figure 1. 24-hours profiles of total triglycerides

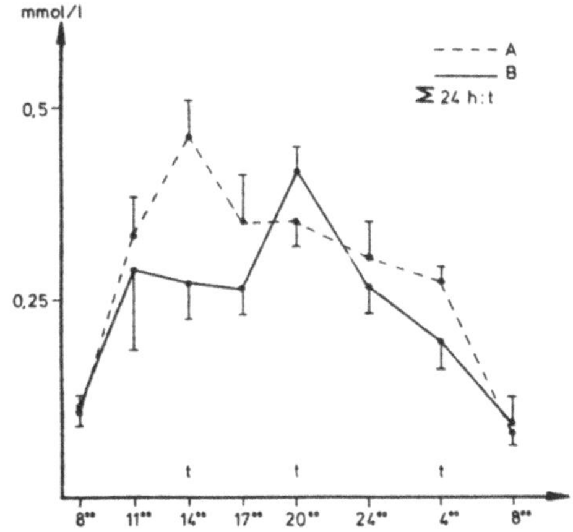

Figure 2. 24-hours profiles of chylomicron triglycerides

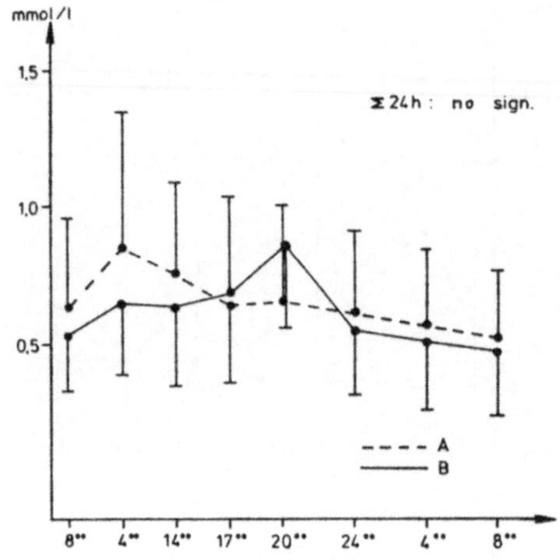

Figure 3. 24-hours profiles of VLDL triglycerides

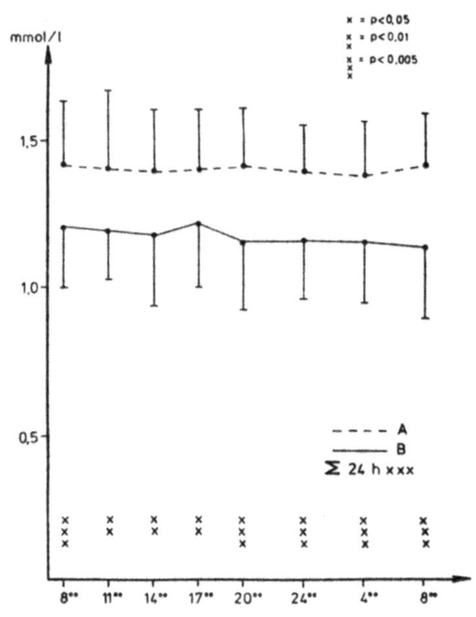

Figure 4. 24-hours profiles of HDL cholesterol

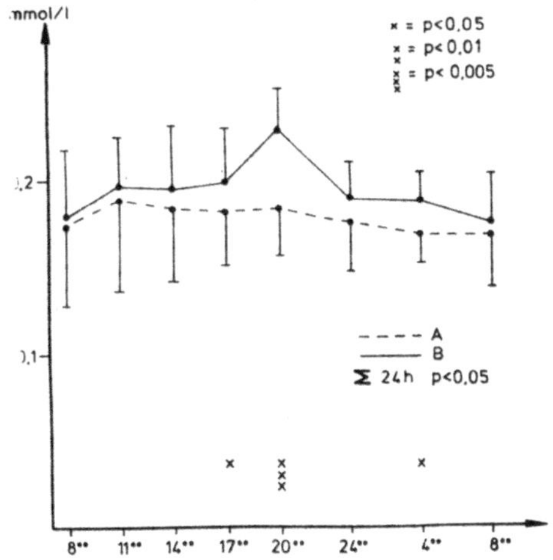

Figure 5. 24-hours profiles of VLDL phospholipids

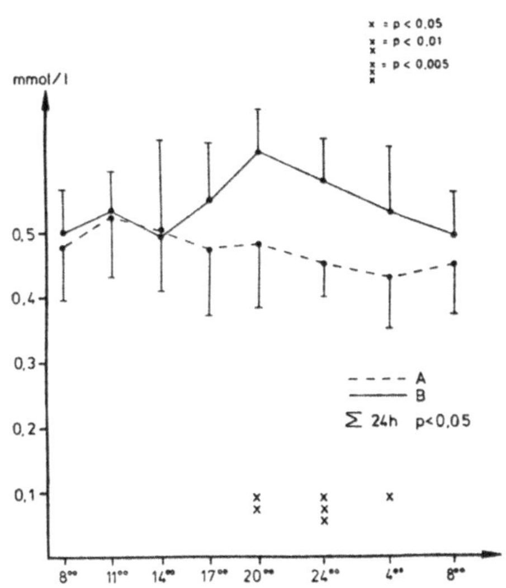

figure 6. 24-hours profiles of LDL phospholipids

Table I. Data of the 12 volunteers

Subject	years	sex	kg	cm	Baseline nutritient intake KJ/Kg and day	plasma chol mmol/L	plasma TG mmol/L
1	25	m	72	176	132	3.18	1.14
2	23	f	54	160	118	4.94	0.86
3	22	m	61	170	128	4.95	1.38
4	30	f	59	168	106	5.14	1.77
5	32	m	70	173	128	5.96	1.43
6	24	m	60	165	135	3.94	0.98
7	22	m	68	180	129	4.25	1.24
8	36	m	60	169	127	4.33	1.09
9	22	m	70	183	133	5.29	2.05
10	23	f	56	162	110	4.36	1.03
11	22	m	70	180	124	4.17	1.14
12	24	f	64	172	118	3.96	1.00

Table II. 24 hour areas of plsama lipid fractions (mmol/L during forms A (15%) and B (40% of total caloric intake as evening meal)

Lipid fraction	n	Form A mean ± SE	Form B mean ± SE	significance
total TG	12	34.41 ± 8.71	33.45 ± 6.45	ϕ
HDL TG	12	6.72 ± 1.57	5.57 ± 1.59	ϕ
LDL TG	12	8.80 ± 1.45	9.10 ± 0.94	ϕ
VLDL TG	12	16.92 ± 6.81	16.19 ± 5.04	ϕ
Chylo TG	4	7.32 ± 0.58	6.14 ± 0.41	ϕ
total chol	12	107.89 ± 14.45	108.27 ± 21.83	ϕ
HDL chol	12	34.05 ± 4.55	28.59 ± 4.99	< 0.01
LDL chol	12	58.32 ± 9.30	59.04 ± 12.85	ϕ
VLDL chol	12	10.54 ± 2.98	11.68 ± 3.19	ϕ
chylo chol	4	4.95 ± 0.62	6.39 ± 1.10	tendency
total PL	12	41.37 ± 6,65	44.15 ± 5.19	ϕ
HDL PL	12	20.95 ± 3.12	20.33 ± 2.91	ϕ
LDL PL	12	11.4. ± 1.88	13.08 ± 1.19	< 0.05
VLDL PL	12	4.26 ± 0.67	4.69 ± 0.49	< 0.05

TG = triglycerides, Chol = cholesterol, PL = phospholipids, HDL = high density lipoprotein, LDL= low density lipoproteins, VLDL = very low density lipoproteins, chylo = chylomicrons showed no difference. Both 8^{00} values of HDL-cholesterol differed significantly during forms A and B ($p<0.005$), with higher values during form A. This difference remained during the whole 24 hours and between the corrected mean values (fig. 4). The mean values of VLDL-PL and LDL-PL were significantly lower with poor evening meal (fig. 5, 6). The diurnal profiles and mean values of total cholesterol, VLDL and LDL cholesterol and HDL phospholipids showed no significant differences between forms A and B (table II), nor did so any parameter compared between females and males.

DISCUSSION

In this study performed in 12 metabolically healthy volunteers, a different diurnal distribution of an isocaloric food intake resulted in a well-known shift in the diurnal peaks in total and chylomicron triglycerides which reflects the actual intake of fat without significant alterations of the mean values over 24 hours. Of higher interest are differences in lipoprotein density classes which correlate - in either sense-with coronary risk, LDL and HDL.

LDL-lipid values remained stable during the 24 hours and without differences between the schedule with poor and rich evening meal. Contrasting to the behaviour of LDL-lipids HDL cholesterol values differed and were significantly lower during the diet with rich evening meal compared to the isocaloric schedule with a poor evening meal.

Increasing of total HDL-cholesterol values have been reported under various conditions like loss of body weight, effective dietetic or drug therapy of hypertriglyceridemia, in runners and premenopausal women, conditions accepted as beneficial in terms of coronary risk. The well-known negative correlation between morning total triglycerides and HDL cholesterol (7) cannot explain our finding of increased HDL-chol in spite of unaltered fasting and 24 hour mean values of total, chylomicron-and VLDL-triglycerides.

At the present state of knowledge, nutritional behaviours which increase HDL cholesterol levels seem to be favourable. The results of this study in normolipemic volunteers would provide a rational basis for a recommendation of a light evening meal-a proverbial golden rule for long.

The discussion of causal relationships between a heavy evening meal and decreased HDL cholesterol values must be highly speculative until the catabolism of triglyceride rich lipoproteins, remnants and intermediate lipoproteins, the participation of lipoproteinlipase and hepatic triglyceride lipase in this process, the origin of HDL in man and a-presumable-indicatorship of HDL levels for an effective clearance of the mentioned lipoproteins is fully understood (8,9,10). Formulated with the appropriate caution, our findings of a delayed diurnal peak in chylomicrons and VLDL-which include chylomicron remnants -, significantly lower phospholipids in VLDL and LDL and decreased HDL cholesterol after heavy evening meals can be interpreted as indicators of a less effective catabolism of probably atherogenic lipoprotein remnants, as has been described in other pathological states (11).

REFERENCES

1. Zilversmit,D.B. 1979. Atherogenesis: A postprandial phenomenon circulation 60: 473 - 485.
2. Roeschlau,P., Bernt, E., Gruber,W. 1974. Enzymatische bestimmung des Gesamtcholesterins im Serum. Z.Klin.Chem.Klin. Biochem.12: 226.
3. Eggstein,M., Kreutz,F.H. 1966. Eine neue Bestimmung der Neutralfette im Blutserum und Gewebe,1. Mitteilung(Prinzip, Durchfuhrung und Besprechung der Methode), Klin.Wschr. 44: 262-267.
4. Schmidt,F.H., Vondahl,K. 1968. Zur Methode der enzymatischen Neutralfettbestimmung in biologischem Material. Z.Klin.Chem. Klin. Biochem.6: 156-159.
5. Bartlett,J.R. 1959. Phosphorus essay in column-chromatography. J.biol.Chem. 234: 466-476.
6. Havel,R.J., Eder, H.A., Bragdon,J.J. 1955. The Distribution and chemical composition of ultracentrifugally separated lipoproteins in human serum. J.Clin.invest. 34: 1345-1353.
7. Schaefer,E.J., Levy,R.I., Anderson,D.W., Danner,R.N., Brewer, H.B.jr., Blackwelder,W.C. 1978. Plasma-triglycerides in regulation of HDL-cholesterol levels. Lancet II: 391-393.
8. Brunzell,J.D., Chait,A., Bierman,E.L. 1978. Pathophysiology of lipoprotein transport. Metabolism, 27: 1109.1127.
9. Tall,A.R., Small,D.M. 1978. Plasma high-density lipoproteins. New Engl. J.Med. 299: 1232-1236.
10. Patsch, J.R., Gotto,A.M., Eisenberg,S., Olivecrona,T. 1978. Formation of high density lipoprotein-like particles during lipolysis of very low density lipoproteins in vitro. Proc. natn. Acad.Sci. USA 75: 4519-4523.
11. Fielding,C.J., Renston,J.P., Fielding,P.E. 1978. Metabolism of cholesterol enriched chylomicrons. Catabolism of triglyceride by lipoprotein lipase of perfused heart and adipose tissues. J.Lipid.Res. 19: 705-711.

ACKNOWLEDGEMENTS

We have to thank Ms. Otte and Ms. Hehnlein for laboratory assistance and the participating students of the medical faculty for their patience.
This work was supported by Public Authority for Science Support.

BIOCHEMICAL ASPECTS OF SOME OBESITIES

P.O. Rodrigues, M.F. de Mequita, M.G.P. de Carvalho,
J. Magalhães, and M.J. Halpern

Dept. of Bioche. of Fac. of Med. Sciences , Center
for Lipid Research from I.N.I.C. and Lab. of Clinical
Chemistry, by grant of the Calouste Gulbenkian Foundation
Campo de Santana 130 - 1100 Lisboa - Portugal

SUMMARY

Diabetes and hyperlipemia are the main metabolic perturbations in Portugal and they related to atherogenic risk.

It's generally accepted that the association of these two metabolic perturbations are pathogenically related.

The profile of the glucose and triglyceride response to OGTT is considered and an attempt to separate these two metabolic perturbations is made.

Hyperlipemias and Diabetes Mellitus are the main metabolic alterations in our country and represent one of the greatest atherogenic risk in adults.

Several authors (1,2,3) refer the high correlation between the atherosclerotic peripherical disease and Hypertriglyceridemia, stating the high incident among Diabetic patients with increased triglycerides on non-Diabetics (4,5).

Diabetic patients with lower members atherosclerosis associated with Fredrickson's Type IV Hyperlipemia are represented in 38%, while the non-Diabetics are only represented in 13% (4).

It's generally accepted (6) that the association of these two methabolic alterations are pathogenically relationed:

1- Hyperlipemia secondary to Diabetes Mellitus
2- Primary Hyperlipemia associated to glucose intolerance and frequently insulin resistant

Among several types of Hyperlipemia, in Portugal, the Type IV is the most common (7).

In the portuguese population (7), the rural one has associated Obesity, Hyperlipemia and Hyperglicemia in almost 50%, while in Lisbon the incidence is lower.

The goal of this work is to study the response to OGTT (50g) of a group of obeses.

METHODS:

Forty five persons were divided in the following groups:

GROUP A - composed by twelve healthy persons with no abnormal laboratory results as a reference group.

GROUP B - composed by twelve obese and diabetic patients.

GROUP C - composed by twelve obese and hyperlipemic patient (Fredrickson's Type IV).

GROUP D - composed by nine obese with abnormal OGTT and hiperlipemia (type IV), divided into 2 sub-groups.

Sub-group D1 composed by four primary Diabetics and *Sub-group* D2 composed by five primary Fredrickson's type IV hiperlipemia.

To each one was performed an oral glucose tolerance test (OGGTT) with a 50 g load of glucose.

The selection criteria was the following:

Healthy persons: Without any clinical signs and normal laboraratory values.

Obese: Any person with more than 10% of the ideal weight according to Lorentz.

Diabetics: With clinical signs, fast glucose and/or post-prandial glucose abnormal and abnormal OGTT.

Fredrickson's Type hyperlipemia: Hypertriglyceridemia and increased preβ-lipoprotein (VLDL) in celulose acetate electrophoresis.

Blood samples were taken at 0, 30, 60, 90 and 120 minutes by venopuncture of the forearm and serum glucose was determined by enzimatic method (glucose-oxidase method described by Werner--Boehringer's Kit) and serum triglyceride determined by fully enzimatic method, modified by Wahlefeld (Boehringer's kit).

The electrophoresis were performed on celulose acetate using a Veronal buffer (8,24g/1) at pH 8.9.

The OGTT interpretation was according to Conn and Fajan's criteria.

RESULTS

Fig. 1 represents the response of glucose and triglyceride to OGTT in control (GROUP A). The maxium values for glucose and triglyceride were both at 30 minutes.

The response of GROUP B was represented in Fig. 2, having the maximum peaks for glucose at 60 minutes and for triglyceride at 30 minutes.

In the Fig. 3, it is represented the response of GROUP C, having the peak values for glucose and triglyceride both at 60 minutes.

SUB-GROUP D1 is represented in Fig 4, having the same peak

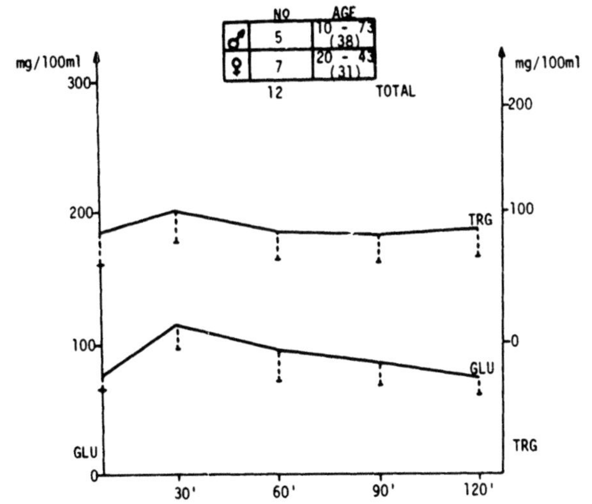

Figure 1. Plasma glucose and triglyceride response (mean ± SD) to OGTT (50g) in 12 healthy controls.

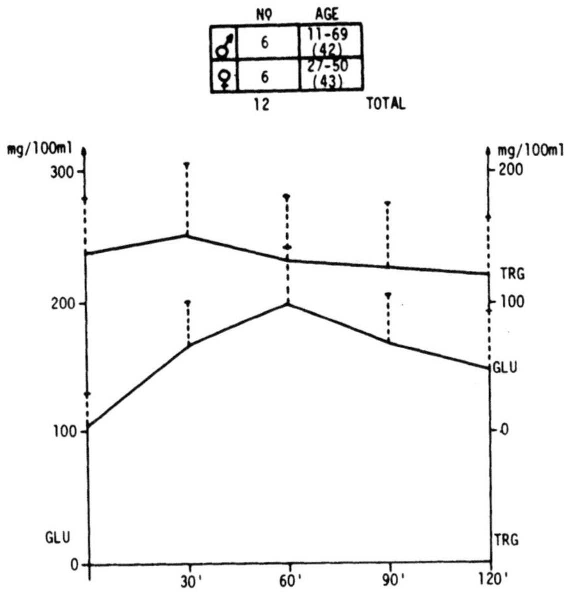

Fig. 2. Plasma glucose and triglyceride response (mean ± SD) to OGTT (50g) in 12 Obese Diabetics patients.

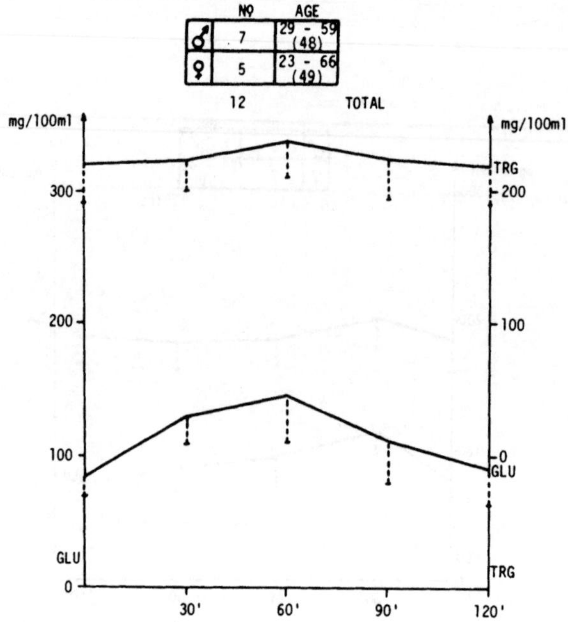

Fig. 3. Plasma glucose and triglyceride response (mean ± SD) to OGTT (50g) in 12 Obese Fredrickson's type IV patients.

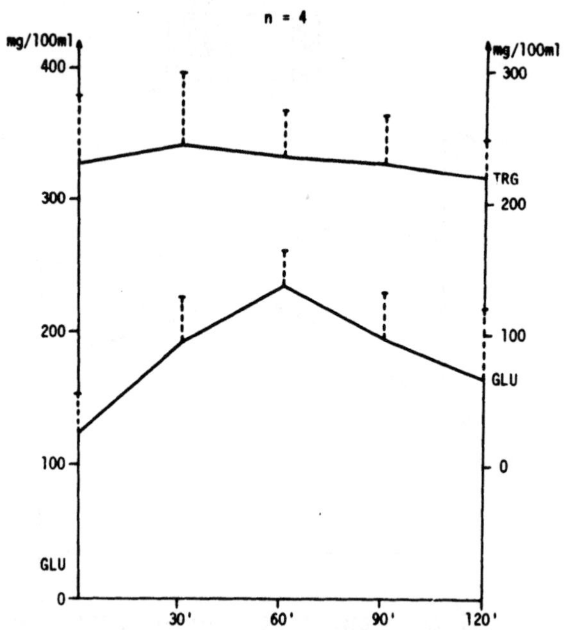

Fig. 4. Plasma glucose and triglyceride response (mean ± SD) to OGTT (50g) in 4 Obese Diabetic and Hyperlipidemic patients (primary Diabetics)

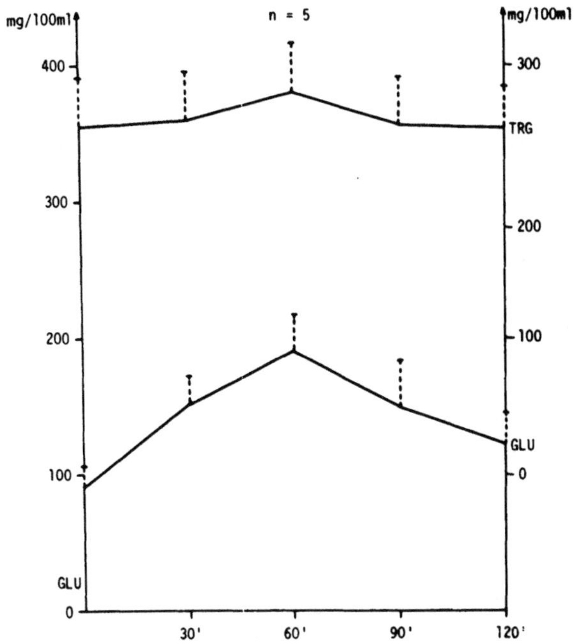

Fig. 5. Plasma glucose and triglyceride response (mean ± SD) to OGTT (50g) in 5 Obese Diabetic and Hyperlipemic patients (primary Hyperlipemics)

OBESE WITH ABNORMAL OGTT
AND
TYPE IV HYPERLIPEMIA

1) Maximum glucose peak at 60 min.	1) Maximum glucose peak at 60min.
2) Maximum triglyceride peak at 30 min.	2) Maximum triglyceride peak at 60 min.
Diabets Mellitus with secondary Type IV hyperlipemia	Primary Type IV Hyperlipemia with secondary carbohydrates intolerance

Fig. 6.

values for glucose and triglyceride as for GROUP B and SUB-GROUP D2 represented in Fig 5 has the same pattern as for the GROUP C.

DISCUSSION:

In GROUP A the OGTT is normal and the peak values for glucose and triglyceride are both at 30 minutes.

In GROUP B composed by obese and diabetic patients, the OGTT is abnormal and the peak value for glucose is at 60 minutes and triglyceride at 30 minutes.

In the group composed by obese hiperlipemics (GROUP C), the OGTT is normal and the peak values are bothe at 60 minutes.

Summarily the diabetic patients have the peak value for glucose at 60 minutes and for triglyceride the peak value is at 30 min., while the hyperlipemic patients have both peaks at 60 min.

In GROUP D composed by obese, hyperlipemic and abnormal OGTT patients, we have verified that in the sub-group D1 the serum glucose and triglyceride values pattern is similar to the GROUP B and the sub-group D2 the pattern is similar to GROUP C.

Some authors (6,8) have described an hyperlipemia test induced by carbohydrates, persisting the increased values of triglycerides beyond the 4th week in the type IV Hyperlipemia.

In conclusion, in an easy and practicle way we can separate the primary type IV hyperlipemia from the type IV Hyperlipemia from the type IV Hyperlipemia secondary to Diabetes Mellitus in Adult Obese, hyperlipemic with abnormal OGTT (Table I).

REFERENCES:

1 - S. Krede S., Kwarstein B. - Acta Chir. Scand 141:335-340, 1975.
2 - Bliss D.P., Kirk CJC, Newall R.G. - Angiology 23:69-75, 1972.
3 - Carlson L.A., Bottinger L.E. - Lancet I:865-868, 1972.
4 - Maria da Graca Paiva de Carvalho, et al.-Momento Médico, Supl. Junho: 3-8, 1980.
5 - Sosenko J.M., Breslow J.L., et al-N. Engl. J. Med. 302: :650-654, 1980.
6 - Jacques Canal - "Hiperlipoproteinémies et Diabetes" in "Problèmes actuels de Biochemie Appliquées". Masson et Cie. 7e serie, 1974.
7 - Halpern, M.J. et al - Afro-Mediteranean Symposium, Milao, 1980.
8 - De Gennes, J.L. - Les Hypercholesterolémies, Gaz. Med. France, 77:6803, 1970.

BENFLUOREX AND LIPID METABOLISM

M.F. Mesquita, M.G. Paiva de Carvalho, P.O. Rodrigues,
J. Magalhães, D. de Matos, and M.J. Halpern

Dept. of Biochemistry and Center for Lipid Research (INIC)
Faculty of Medical Sciences UNL
Campo de Santana 130, P-1100 Lisboa - Portugal

In the western countries it is often observed the association of obesity with hyperglyceridemia, hyperuricemy and decreased glucose tolerance, described by Crepaldi as the Plurimetabolic syndrome. In this situation there are an accumulation of several risk factors in the same subject.

The best treatment of this situation is the diet but when diet is not enough we need to use drugs therapeutic.

In this work we shall describe our experience with the action of Benfluorex (x) in 25 patients submited with an hypocaloric diet associated to the administration of unsaturated fatty acids.

The results before the treatment are shown in tables I and II.

Table I. Results before treatment (mmols/l)

	$\bar{n}$	o
Triglycerides	28, 01	5,45
Cholesterol	6,32	1,04
Uric acid	0,35	0,06
HDL cholesterol	0,99	0,18
Glucose	4,79	1,10

Table II. Tolerance glucose test

Minutes	Glucose	σ
0	4,79	1,10
30	7,48	1,65
60	8,80	1,65
90	6,93	1,65
120	5,78	1,65

Benfluorex was given as shown in table III.
The results are shown in tables IV and V.
In table VI we sumarize the action of Benfluorex.
We can conclude that benfluorex is a drug that can be used in plurimetabolic syndrome as a whole.

Table III. Benfluorex administration

Week	Breakfast	Lunch	Diner
1st	-	-	+
2st	-	+	+
3d and followings	+	+	+

Table IV. Effects of Benfluorex (mmols/l)

	Before	after
Triglycerides	28,01 ±5,45	16,90 ±2,18
Cholesterol	6,32 ±1,04	5,88 ±0,78
Uric acid	0,35 ±0,06	0,33 ±0,06
HDL cholesterol	0,99 ±0,80	1,12 ±0,21
Glucose	4,79 ±1,10	4,7 ±0,55

Table V. Benfluorex and Glucose tolerance test

Minutes	Before	After
0	4,79	4,07
	±1.10	±0,55
30	7,48	6,99
	±1,68	±1,10
60	8,80	7,6
	±1,65	±1,10
90	6,93	6,27
	±1,05	±1,10
120	5,78	5,12
	±1,65	±0,55

Table VI. Results of Benfluorex administration

Parameter	% of alteration
Triglycerides	- 39
Cholesterol	- 7
Uric acid	- 7
HDL cholesterol	+ 14
Glucose	- 9
Glucose tolerance	Increased

ISOLATION AND PARTIAL CHARACTERIZATION OF Lp-A-I AND Lp-A FROM HDL (HIGH DENSITY LIPOPROTEINS) OF HUMAN SERUM

P-L. Norfeldt, S.-O. Olofsson, G. Fager, and G. Bondjers

The Arterial Biology Group, Department of Medical Bichemistry and Medicine I, University of Goteborg Goteborg, Sweden

INTRODUCTION

The concept of lipoprotein families (1) defines the lipoprotein family as the base for the lipoprotein system. The lipoprotein family is characterized by a single apolipoprotein as protein moiety. The following lipoprotein families have been identified: Lp-A-I, Lp-A, Lp-B, Lp-C, Lp-D, Lp-E and Lp-F. This paper deals with the isolation and partial characterization of the lipoprotein families A-I (Lp-A-I) and A (Lp-A) from the high density lipoproteins of human serum.

MATERIAL AND METHODS

High density lipoproteins (HDL) were isolated by sequential preparative ultracentrifugation between densities 1.063 and 1.2.. HDL were washed by repeated recentrifugation at the upper density until free of albumin as judged from immuno-double diffusion.

CHROMATOGRAPHIC PROCEDURES

Ion-exchange chromatography was performed on QAE-Sephadex A 50 (Pharmacia Fine Chemicals, Uppsala, Sweden) column (20x1.5cm). The column was equilibrated with 0.07 M phosphate buffer, pH 6.0 with 0.05% sodium EDTA and 0.05% sodium azide (the start buffer). The column was eluted with the start buffer followed by a sodium chloride gradient (0 t0 0,35 M) in the start buffer (100+100 ml).

Hydroxyl apatite (Bio Gel HTP, Bio Rad, Richmond, Calif., USA) column chromatography was performed on a 20x1.5 cm column. The column was eluted with phosphate buffer of stepwise increasing molarity: 0.001, 0.01, 0.05 or 0.06 and 0.15 or 0.65 M.

Gel chromatography was performed on a Bio Gel A-5 M (Bio Rad, Richmond, Calif., USA) column (250x1.5 cm) equilibrated with 0.01 M Tris, pH 7.5 with 0.5 M sodium chloride and 0.015 sodium azide.

The column was standardized with the following standards of known diameter: Thyroglobulin, LDL-2, HDL-2 and HDL-3.

Electrophoresis of apolipoprotein was carried out on 7% polyacrylamide gels containning urea at basic (2) and acidic (3) pH and on NaDodSO, (sodium dodecyl sulphate) containing polyacrylamide gels (4).

ISOELECTRIC FOCUSING

Flat bed isoelectric focusing was carried out on 1% IEF--agarose (Pharmacia Fine Chemicals, Uppsala, sweden) gels containing 12% sorbitol and 2% Ampholine (LKB, Bromma, Sweden). The procedure was carried out at a constant power of 5W. The pH gradient was measured with a surface electrode (LKB, Bromma, Sweden).

ELECTRON MICROSCOPY

Lipoprotein were negatively stained with phosphotungstic acid (5) and examined in a Jeol JSM 100 CX transmission electron microscope (Jeol, Tokyo, Japan) with an accelerating voltage of 80 kV.

Density gradient centrifugation was carried out in a Superspeed 65 preparative ultracentrifuge utilizing a 8x35 ml rotor. The density gradient was formed by mixing equal volumes of 0.001 M phosphate buffer, pH 6.0 with 0.05% sodium EDTA and sodium azide adjusted to densities 1.050 and 1,25 by solide potassium bromide in a gradient mixer. The gradient was unloaded with a gradient extraction unit(MSE, West Sussex, England).

Amino acid analysis was performed with the methods used earlier (6).

Protein quantification was performed with the method of Lowry et al (7).

Quantification of apo A-I and A-II were performed with electroimmuno assay (8,9).

Lipid quantification was performed with the methods used earlier (6).

IMMUNO-CHEMICAL METHODS

Preparation of apolipoproteins were performed with the methods used earlier (10,11).

Crossed immunoelectrophoresis was performed with the method of Weeke (12). Immuno double diffusion was performed in 1% agar gels.

RESULTS

Isolation of Lp-A-I

HDL was chromatographed on a QAE-Sephadex column. An unretained fraction was obtained. This fraction was further purified by hydroxyl apatite column chromatography. The fraction eluted from this column with 0.06 M phosphate buffer was chromatographed once more on QAE-Sephadex A50. The unretained fraction obtained during this chromatography contained Lp-A-I.

CHARACTERIZATION OF Lp-A-I

The protein moiety of Lp-A-I contained only Apo-A-I as judged from electrophoresis on basic and acidic polyacrylamide gels containing urea and on $NaDodSO_4$ containing polyacrylamide gels, amino acid composition and immuno double diffusion.

Lp-A-I appeared as spherical particles with a diameter similar to HDL on electron microscopy. The diameter was estimated to be 95 Å from gel chromatography.

Lp-A-I banded in a density gradient in the HDL-2 region, with a maximum at d 1.095. No Apo A-I reacting material could be detected in the density region greater than 1.135.

Lp-A-I migrated as one precipitin arc on crossed immuno electrophoresis. The migration correponded to that of HDL. On isoelectric focusing Lp-A-I appeared as multiple bands in the pH region 5.05 to 5.50.

Phosphatidyl choline and cholesterol esters were the quantitatively dominating lipids.

ISOLATION OF Lp-A

The HDL fraction retained by the QAE-sephadex A 50 column was eluted with a sodium chloride gradient and chromatographed on a hydroxylapatite column. The fraction eluted from this column with 0.05 M phosphate buffer was further purified by chromatography on a anti-D containing immunoadsorber. The unretained fraction from this column contained Lp-A.

CHARACTERIZATION OF Lp-A

The protein moiety of Lp-A contained Apo-A-I and A-II as judged from electrophoresis on basic and acidic polyacrylamide gels containing urea and from immuno double diffusion. The result from immuno double diffusion indicated that the two polypeptides occurred on the same lipoprotein particles. The ratio between Apo-A-I and A-II was estimated to be 1.3 $\pm$ 0.3 (n=4) from the amino acid composition. Quantification of apo-A-I and A-II with electro-immuno assay gave a ratio of 1.4 $\pm$ 0.2 (n=4). Lp-A appeared as spherical particles similar to HDL. The diameter estimated from gel chromatography was 72 Å .

Lp-A banded in a density gradient in the HDL-2 region with a peak at 1.090 g/ml. No apo-A-I or A-II reacting material could be detected in the density region greater than 1.140 g/ml after density gradient centrifugations of Lp-A. Apo-A-I and A-II had the same distribution in the density gradient and the ratio between the two polypeptides was constant.

Lp-A appeared as one precipitin arc on crossed immuno electrophoresis with antiserum to apo-A-I or A-II. The migration was similar to that of HDL. The precipitin arc formed between Lp-A and anti-A-I and Lp-A and anti-A-II occurred on the same place on crossed immuno electrophoresis.

On isoelectric focusing Lp-A appeared as two protein staining bands. Phosphatidyl choline and cholesterol esters were the quantitatively dominating lipids.

DISCUSSION

The present paper presents isolation procedures for the lipoprotein families Lp-A-I and Lp-A.

The presence of lipoprotein particles containing apo-A-I as single protein moiety is suggested from several observations (13,14,15). It has, however, been reported that apo-A-I could be detached from HDL and obtained in lipid free form (16,17,18).

The observation in this paper indicate that Lp-A-I consists of spherical particles with the physical chemical properties of a high density lipoprotein. Our data indicate that the apo-A-I and A-II occur on the same particles within Lp-A thus forming an apo-A. Our data on Lp-A confirms findings by other authors (19).

ACKNOWLEDGEMENTS

This study was supported by grants from the Swedish Medical Research Council (no 4531), ths Swedish Oleo-Margarine Foundation for Nutritional Research, the Swedish National Association against Heart and Lung Diseases and the University of Goteborg.

We thank Anita Magnusson, Margit Wettesten, Gun Olsson, Ulla-Britt Rignell and Aire Lidell for skiiful technical assistance and Agneta Ladstrom for typing the manuscript.

REFERENCES

1. Alaupovic, P. (1972) Proteides Biol. Fluids Proc. Colloq. 19, 9-19
2. Davis, B.J. (1964). Ann. N.Y. Acad. Sci. 121, 303-427.
3. Reisfeldt, R.A., Lewis, U.J. & Williams, D.E. (1962). Nature (London) 195, 281-283.
4. Weber, K. & Osborn, M. (1969). J. Biol. Chem. 244,4406--4412.
5. Hayatt, M.A. (1970). Principles and technique of electron microscopy. Biol. Applications 3,5.
6. Olofsson, S.-O. & Gustafson, A. (1976), Scand. J. Clin. Lab. Invest. 36, 67-74.
7. Lowry, O.H., Rosenbrough, N.J., Farr, A.L.,& Randall, R.I. (1951). J. Biol. Chem. 20, 151-157.
8. Fager, G., Wiklund, O., Olofsson, S.O., Norfeldt, P.-I., Vedin, A. & Bondjers, G. (1980), Scand. J. Clin. Lab. Invest. 40, 451-460.
9. Fager, G., Wiklund. O., Olofsson, S.-O., Norfeldt, P.I., Wilhelmsen, L. & Bondjers, G. (1979), Artery 6, 188-204.
10. Olofsson, S.-O., McConathy, W.J. & Alaupovic. P. (1978), Biochemistry 17, 1032-1036.
11. Olofsson, S.-O., Bostrom, K., Svanberg, U. & Bondjers, G. (1980). Biochemistry 19, 1059-1064.
12. Weeke, B. (1973) p. 47-56 in Axelsen, N.H., Kroll,J. &

Weeke, B. eds.A manual of quantitative immunoelectrophoresis. Methods and applications. Universitets forlages Oslo, Norway.
13. Albers, J.J. & Aladjem, F. (1971), Biochemistry 10, 3436--3442.
14. Olofsson, S.O., Gustafson, A. & Svanberg, U. (1975), Scand. J. Clin. Lab. Invest. 35, 363-371.
15. Patsch, W., Schonfeld, G., Gotto, Jr, A.M. & Patsch, J.R. (1980). J. Biol. Chem. 255, 3178-3185.
16. Lagocki, P.A. & Scanu, A.M. (1980). J. Biol. Chem. 255, 3701-3706.
17. Olofsson, S.-O., Fager. G & Gustafson, A. (1976), Scand. J. Clin. Lab. Invest. 36, 481-488.
18. Nichols, A.V., Lux. S., Forte. T. & Levy. R.I. (1972). Biochim. Biophys. Acta 270. 132-148.
19. Kostner, G. & Alaupovic. P. (1972). Biochemistry 11. 3419-3428.

TYPE IV HYPERLIPEMIA

M.J. Halpern

Center for Lipid Research
Dept. of Biochem. , Fac. Medical Sciences
Compo De Santana, 130 - 1100 Lisbon-Portugal

DEFINITION

Type Iv is an hyperlipemia characterized by an increase of VLDL (prē-beta lipoproteins). This increase is generally associated with a triglycerides increase. However, there are a special type (latent type IV) with normal triglycerides.

SCREENING

Using the heparin - sodium chloride it is possible to separate a population in normal and probably hyperlipemic. With this test it is not possible to screen any particular type of hyperlipemia. The association of this test with the sodium laurylsulphate test or eventually the utilization of HPTLC (1,2) afford this creening. Although the heparin-sodium chloride can be a very good general screening method in epidemiological studies we think that the creening cannot remplace typification in the creened population and for that reason specific screening tests as sodium laurylsulphate are useless.

TYPIFICATION

Typification of type IV is done by cholesterol and triglyceride determination and lipoprotein electrophoresis. However it is useless to do an electrophoresis to a patient with increased triglycerides with normal cholesterol because there is not possible a diagnosis different from type IV.

SPECIAL TYPES

There was been described special types different classical type IV.

1. Type IV b - In same patients it is possible to see in electrophoresis a double pré-beta. One of the bands as the same migration of pré-beta and the other migrates slowly and it is named extra pré beta, pré-betal or slow pré-beta (3-7). De Gennes studied the charcteristics slow pré-beta by ultracentrifugation and thinks this presence of a sign of bad prognosis. By analogy with IIb type De Gennes proposes for this situation the designation of IVb type.

2. Latent type IV - We described (8-11) increase pré-beta with normal triglycerides situation we named IV or IIb normotriglyceridemic types.

This situation was described by Loeper and col (12) as a pathological state and by Fredrickson (13) as a normal situation caused by a polymorphic form of beta-lipoproteins that he named sinking beta later identified as Lp (a).

Our results let us to think that in some cases the sinking beta cannot explain this finding.

1. Some of these results were observed in families of type IV hyperlipemies.

2. In a study of a family in New Foundland, Angewill described a member of the family only with an increased pré-beta (14).

3. In 40 women in the first days of post-partum we observed 37 type IV and 3 with increased pré-beta.

4. After treatment in same cases triglycerides fall to normal values but pré-beta remain increased.

In order to understand this problem we made the test proposed by de Gennes (15-16) - determination of triglycerides after 8 hours fasting. De Gennes found that at 8 hours fasting normal subjects and type IIa patients show normal values and type IV and IIb increased values. With this test we find two kinds of results (table I): (I) increase < 0,4 mmol/l (2) increase >0,4. We consider the second as mild forms of types IV or IIb for what we proposed the name of latent IV or IIb types.

It is very interesting the evaluation of a type IIb with treatment (table II).

However this 8 hours fasting does not afford a good timetable for laboratory determination. In order to find another timing we determined triglycerides 0, 1, 3, 5 and 7 hours after lunch and we found that normal subjects did not present any increased of triglycerides at the 3 hours samples. For that reason we conclude that for the characterization of these normotriglyceridemic types was enough the 3 hours fasting (17).

EPIDEMIOLOGY

Type IV is the most important hyperlipemia in the Portuguese population. In the Table III we reproduce results of two studies one realized in a factory of Lisbon and other in arural population (18).

Goulão and al. (19) described similar results in Coimbra region.

Table I. Ge Gennes test in some normotriglyceridemic type IV

Hours of fasting	Triglycerides (mmol/l)					
12	1,60	1,51	1.42	1,35	1,60	1,74
8	1,83	1,54	1,65	1,96	2,46	2,46
Conclusion	Normal (Sinking beta)			Pathological (latent, type IV)		

Table II. Evolution of type IIb

	Cholesterol (mmol/l)	Triglycerides (mmol/l)	Electrophoretic patern
Before treatment	7,74	6,30	IIb
After treatment			
12 hours fasting	8,26	3,79	IIb
8 hours fasting	8,26	4,70	IIb

Table III. Prevalence of hyperlipemies in % in two different population

Patern	Lisbon (n=139)	Rural population (n=152)
Normal	71,9	59,9
IIa	2,2	4,0
IIb	2,2	5,3
IV	23,7	30,9

RELATIONSHIP WITH GLUCIDES

It is well known that carbohydrates can increase triglycerides levels. However the kind of ingested glucides is not indifferent. Observing the results of table IV, taken from the experiences of Krithevsky (20), we can conclude that in rats all the glucides essayed increase triglycerides but that increase is greater with fructose and sucrose.

The sucrose can be attributed to fructose.

The same effect was described in man.

There exists an individual sensibility. We could demonstrate that normalipemic subjects do not show important increases with a load of 50 gr of fructose versus type IV patients in which we observed increases (21-22).

A load of glucose does not provoke such increase.

RELATIONSHIP WITH URIC ACID

Several factors call the attention to the relationship between hypertriglyceridemia and hyperlipemia.

(1) Hyperuricemy is frequent in illnesses associated with hyperlipemies-cerebral vascular accidents, coronary disease, arteriopathies.

(2) Gout is often associated with increased triglycerides.

(3) Medicaments that decreases uric acid decrease at same time triglycerides - is the case of halophenate and sulphinpyrazol.

Some authors describe an association between hyperlipemies and hyperuricemy. De Gennes and all studying 920 subjects described 60% of hyperuricemy on hypertrigliceridemy, 30% in patients only with hypercolesterolemy and 18% in normal people.

In table V we show our results (21, 23-26).

We studied several correlations between these parameters (table V).

For the correlation triglycerides-cholesterol we calculate the correlation Y=0,0048,4x4,771.

We can explain this correlation saying that the increase of uric acid is caused by the increase of triglycerides.

This correlation is shown by the decrease of uric acid in type IV patients by hypoglucid diets (but sometimes hyperproteic). In table VII we give an example.

The fructose load increase at the same time triglycerides and uric acid (table VIII).

However this relationship is true only in hyperlipemia. In gout and pré-gout the fructose load increases only uric acid (table IX).

PLURYMETABOLIC SYNDROME

The association of some nutritional errors with a genetic terrain has created a specific syndrome of western civilization the plurymetabolic syndrome characterized by obesity, low glucose tolerance (or fasting hyperglicemia), hypertrigliceridemia,

Table IV. Action of several glucides on serum triglycerides

One year of alimentation with	Triglycerides mmol/l
Control	0,83 ± 0,034
Glucose	1,22 ± 0,034
Starch	1,24 ± 0,036
Fructose	1,46 ± 0,15
Sucrose	1,37 ± 0,13

Table V. Uric acid and hyperlipemic

Pattern	Uric acid (% of cases)			
	300	300-360	360-420	420 μmol/l
Normal	57,3	27,6	15,7	12,0
IIa	5,1	8,0	4,3	5,3
IIb	10,8	13,8	18,6	18,7
III	0,1	0	0	0
IV	26,7	40,5	60,0	64,0
V	0	1,1	1,4	0
N	157	87	70	75

Table VI. Correlations

	Correlations
Cholesterol-Uric acid	No
Uric acid-cholesterol	No
Uric acid-triglycerides	No
Triglycerides-uric acid	Yes

TABLE VII. Uric acid and hypoglucidic diet

Month of diet	Triglycerides mmol/l	Uric acid µmol/l
0	3,96	492
4	2,46	300

Table VIII. Fructose load in a type IV patient

	0'	60'	90'	180'
Triglycerides	246	2,80	2,75	2,57
Uric acid	300	564	452	

Table IX. Fructose load in pré-gout

	0'	30'	60'	90'	120'
Triglycerides	2,70	2,64	2,54	2,46	2,61
Uric acid	324	372	372	390	366

hyperinsulinism and polycythemia (28-29)

Sometimes there are problems of diferencial diagnosis with diabetic obesity. We demonstrate that it is possible this diferencial diagnsosis observing the timing of the maximum peaks of glucose and triglycerides or an oral load of 50 gr. of glucose. (table X).

RELATIONSHIP WITH ALCOHOL

To study alcohol dependence of type IV patients we proposed a load test with 0,41. of wine (13º) with determination at 0, 15, 30, 60 and 120 m. In normal subjects we did not observed increases of triglycerides. In some type IV we found the same results but in the most part of cases an increase that continue after the 120 minutes sample. Therefore we conclude that we can divide type IV in alcohol sensitive and in no alcohol sensitive (30).

Table X. Timing of maximum peak or glucose and triglycerides in a glucose load test

	Glucose	Triglycerides
Normal	30	30
Diabetic	60	30
Type IV	60	60

RELATIONSHIP WITH MAGNESIUM

There was described an inverse relationship between hardness of water and coronary disease. For that reason we studied the serum magnesium on hyperlipemies. We found a signification decrease in hyperlipemies but only on types IV and IIb (31)

REFERENCES

1. Kupke, I. and Engner, S - J. Chromat. 146:261 (1978)
2. Halpern, M.J. - Rev. Port. Bioq. Apl. - In Press
3. Dhalen, G. et all. - Acta Med. Scand. Supl. 531:11 (1972)
4. Dhalen, G. - Acta Med. Scand. Suppl. 570 (1974)
5. Cloarec, M. et all. - Ann. med. int. 125:737 (1974)
6. De Gennes, J.L., et all. - Ather. 27:493 (1977)
7. Pagan, A., Zanetti, G. and Guarini, P. - Ather. 35:351 (1980)
8. Halpern, M.J. and Miguel., M.J.P. - Conference on serum lipoproteins (Graz, 1973)
9. Halpern, M.J. - Nouvelle Press Med. 2:3124., (1973)
10. Halpern, M.J. - Acta Diab. Lat., 11:173., (1973)
11. Halpern, M.J. - Mom. Med., 219:43 (1977)
12. Loeper, J. et all., - Ann. Med. Int., 122:551., (1971)
13. Frederickson, D.S. - Modern concepts of cardiovascular disease., 41:31., (1972)
14. Angilvel, L. - Can. Med. Ass. J. 108:60 (1973)
15. De Gennes and Truffert, J., - C.R. Acad. Sci., 273:2312., (1971).
16. De Gennes, J.L. Turpin, G. and Truffert, J., - Nouv. Press. Med. 1:1627 (1972)
17. Halpern, M.J. - Proc. 2nd Int. Congr. Clin. Bioch. (Prague, 1973)
18. Miguel, M.J.P. and Halpern, M.J. - Ann. Biol.Clin., 3:145 (1973)
19. Goulão, M.J. and Cardoso, S.M. - Rev. Port. Bioq. Appl. 1:25 (1978)
20. Kritchvensk, D. and Tepper, S.A. - Nutr. Rep. Int., 8:163, (1973)

21. Halpern, M.J., and Miguel, M.J.P. - Lancet, I, 670, (1973)
22. Halpern, M.J., Am. J. Nutr. 26:687, (1973)
23. Halpern, M.J.,- Am. J. Cardiol. 95:540 (1973)
24. Halpern, M.J.,- Acta Med. Scand. 202:335., (1977)
25. Halpern, M.J.,- Mom. Med., 219:43 (1977)
26. Halpern, M.J. and Morais Sarmento - Proc. I Int. Col. Hyperlipemia (Lisboa, 1974)
27. Halpern, M.J. - Proc. I Afro Medit. Colloquium Clin. Chem., (Milan, 1980)
28. Halpern, M.J. - Med. Cir. 1:5, (1981)
29. Halpern, M.J. - Proc. 3d. Europ. Congr. Clin. Chem. (Brighton, 1979)
30. Halpern, M.J. - In Deficiencia de Magnésio, ed. Aurélio Rapado (Fundacion Jimenez Diaz, Madrid, 1981)

TESTING HUMAN LIPOPROTEINS FOR MUTAGENIC ACTIVITY IN THE SALMONELLA/MICROSOME ASSAY

J. Rueff and M.J. Halpern

Dept. of Biochemistry and Centre for Lipid Research (INIC)
Faculty of Medical Sciences UNL
Campo Santana 130, P-1100 Lisboa - Portugal

The hypotheses and analogies in themselves have no value in scientific investigation except to the extent that they function as entering wedges for further investigation.

R. VIRCHOW

("Standpoints in scientific medicine", 1847)

INTRODUCTION

In ancient Rome mean human lifespan was about 22 years whereas nowadays in USA it is about 74 years. Although this dramatic difference actually reflects a substantial improvement in general health care measures it also brought about a huge and increasing concern on the so-called "diseases of aging", namely cardiovascular diseases and cancer as these two conditions represent the leading causes of death and disability in developed countries as well as the major challenges facing medical research.

Thus, when discussing lipid pathology a main point of concern is its relationship with atherosclerotic cardiovascular diseases. Atherosclerosis is usually discribed as a multiple factorial disease. Numerous factors, besides hyperlipemia, like cigarette smoking, high blood pressure, various genetic diseases and many others have all been linked to the development of the disease. However, correlations (no matter how good they are) do not constitute proofs of mechanism. As a matter of fact, a thorough causal relationship or even a clear picture of the problem is yet to be obtained. This implies, from the point of view of fundamental research, the choice of experimental

Abbreviations: B(a)P: benzo(a)pyrene; DMBA: 7,12-dimethylbenz (a,h) anthracene; DMSO: dimethylsulfoxide.

approaches within the framework of conceptual models. For the sake of simplicity, let us set one such model in a thesis-like form, which may stay as follows: "At least some types of atherosclerotic plaques in human behave as smooth muscle tumors of the arterial wall and may hence have as possible causes carcinogenic agents". We ought to stress that such a thesis is far from beeing demonstrated. It is possible, however, to gather some data supporting it.

The first datum which allowed the proposal of such a model came from the observation that atherosclerotic plaques have, like tumors, a monoclonal character[1,2]. Assuming that the nature of the two processes could be similar, then one could expect carcinogenic agents to enhance the development of atherosclerosis. This hypothesis has been demonstrated using a chicken model[3]. The results show that atherosclerotic lesions of the abdominal aorta occur more frequently and are larger in DMBA - and B(a)P-treated animals and are proliferative in character.

Regarding the role of plasma lipoproteins in the process, an interesting contribution came from "in vitro" incubation studies with B(a)P which showed that lipoproteins are responsible for greater than 95% uptake of B(a)P by total plasma[4]. This observation rises the question of whether lipoproteins may function as preferential carcinogen transporters "in vivo". As an approach to study the question we have tested lipoprotein fractions (VLDL, LDL and HDL) for mutagenic activity in the Ames Salmonella/microsome assay. We have studied a random group of 20 individuals including normolipemic and hyperlipemic subjects.

MATERIALS AND METHODS

Lipoproteins fractions

Blood was drawn from all 20 subjects studied after an overnight fast. Total serum cholesterol and triglyceride as well as cholesterol and triglyceride of lipoproteins were determined by fully enzymatic methods. Lipoproteins were isolated by sequential ultracentrifugal flotation techniques.

Salmonella/microsome assay

Mutagenicity assays were carried out following closely the published procedure of Ames <u>et al</u>[5].except that 2.5 ml instead of 2 ml of top agar were added to Vogel-Bonner plates. Lipoprotein fractions were always tested dissolved in DMSO in the presence and in the absence of liver enzymes (S-9 fraction) prepared from rats induced with Aroclor 1254. A ratio of 0.1 for S-9 fraction/S-9 Mix was used throughout these experiments.

Strains TA 98 and TA 100, obtained through the courtesy of Prof. B.N. Ames, were used as tester strains.

The activity of S-9 Mix was checked in each experiment by measuring the mutagenicity of B(a)P dissolved in DMSO. Strains were also checked in each experiment for <u>amp</u>^{r}, <u>rfa</u> and <u>uvr B</u>. <u>Rfa</u> was checked on McConkey plates on which <u>rfa</u> strains are growth deficient because of bile salts. The presence of uvr B was checked using the gal character as an indicator on EMB gal plates (since gal and uvr B are included in the same deletion. The controls which were run to obtain the spontaneous background mutation rate included the solvent (DMSO) and the salt solutions used to adjust density for ultracentrifugation.

RESULTS AND DISCUSSION

Three out of the 20 individuals'random group studied presented significant mutagenic activity but restricted to the VLDL fraction when tested with microsomal enzymes activation. Results were considered positive only when an enhancement of at least 2 times the number of spontaneous revertants was obtained along with a dose--response curve. Typical dose-response curves of VLDL fraction from subjects A and B together with the one for B(a)P are presented in Fig.1 for strain TA 98, in a semi-log scale with the amount of VLDL used per plate in µl and that of B(a)P in µg. Spontaneous revertants have not been subtracted (they were always less than 60 per plate). However, a more convenient way to express and compare the results, is to calculate the slope of a linear dose-response plote. Fig 2 is an example of such a plot for VLDL fraction from subjects A and B tested with activation in stains TA 98 and TA 100.

Mutagenic activity of VLDL fraction from the three positive individuals (expressed as rev/µl of VLDL) is presented in Table 1 along with cholesterol and triglyceride concentrations in the serum and in the VLDL fractions. As can be seen from Table 1, there is no correlation between mutagenic activity of VLDL and total cholesterol levels. Yet, in what concerns total triglyceride, VLDL cholesterol

Table I. Mutagenic activity and concentrations of cholesterol and triglyceride in serum and VLDL from individuals A, B and C.

Individual	Serum Chol. mg%	Serum TG mg%	VLDL Chol. mg%	VLDL TG mg%	Mut. Activ. rev/µl
A	265	65	25	30	2,75
B	185	100	35	70	12,50
C	325	280	135	230	14,80

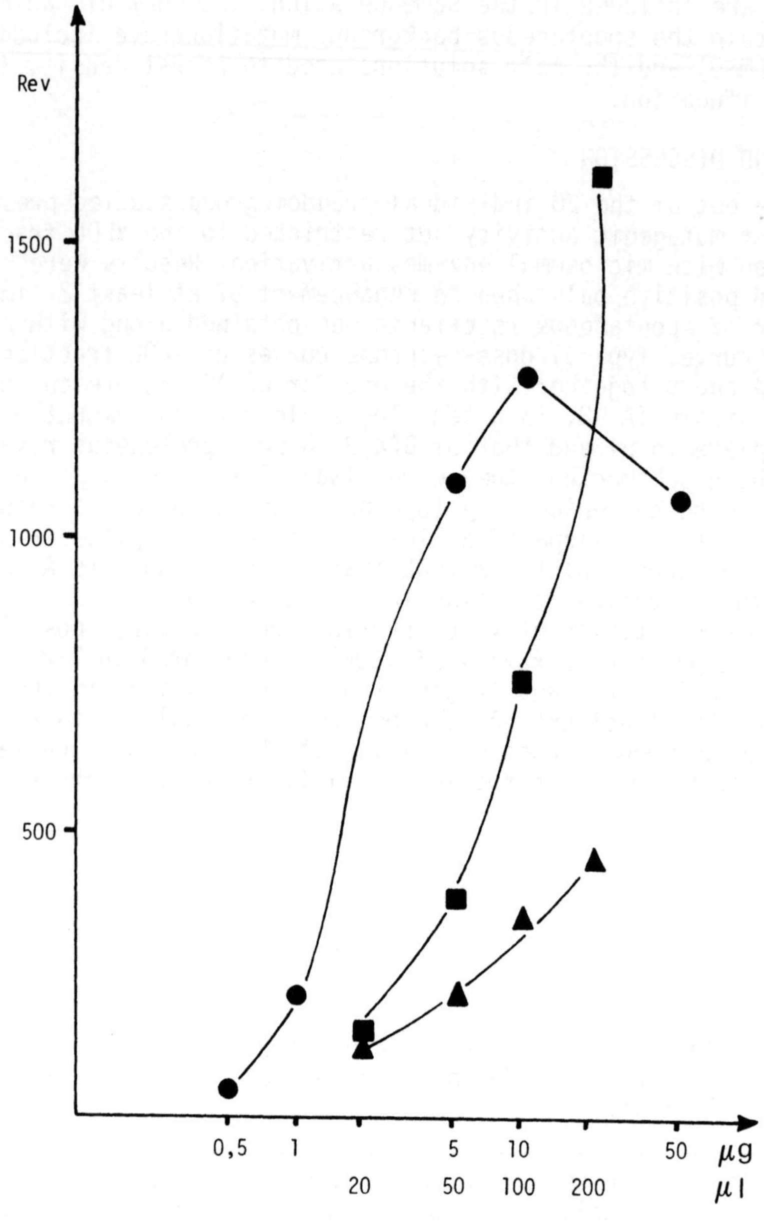

Figure 1. Reversion of TA 98 with B (a) P (●) and VLDL from subjects A (▲) and B (■). The procedure was carried out as described in Materials and Methods in the presence of liver enzymes.

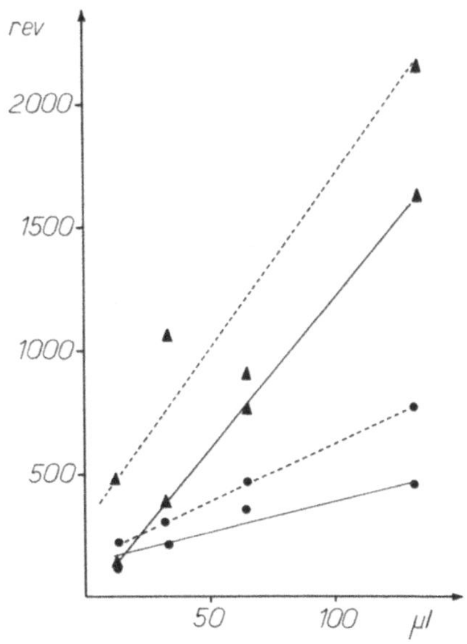

Figure 2. Dose-response effect of VLDL from subjects A (●) and B (▲) in strains TA 100 (dashed line) and TA 98 with metabolic activation.

and VLDL triglyceride concentrations, the specific mutagenic activity of VLDL seems to increase with increasing levels of these three parameters (although in a non-linear fashion). This is not a surprising observation if one thinks of VLDL mutagenic activity dependent upon its capacity to uptake and transport mutagens in circulation which would increase in paralell with the amount of lipids carried by that lipoprotein fraction. Knowing that genetic factors seem to play a relatively great role in the variation of VLDL levels, we may further speculate that environmental apolar compounds (other than apolar lipids) can also enhance the synthesis of VLDL. Supporting this hypothesis is the fact that cigarrete / smoking is associated with 20 to 30 per cent increase in VLDL in both sexes[6].

It remains to be demonstrated whether the absence of mutagenicity associated with LDL and HDL is simply a technical shortcoming of our procedure or can be explained by any metabolic reason.

In the first hypothesis we could think that, at least LDL, have a higher affinity for apolar compounds than VLDL and thus stronger extraction methods must be used instead of simply using DMSO as a solvent. However a metabolic explanation may not be ruled out. Cellular uptake of apolar xenobiotics seems to involve simple partitioning of these compounds between the plasma lipoproteins and the cell membrane lipid volume without the need for a specific process involving cell receptors for LDL. This is apparently the conclusion which can be drawn from the study by Remsen and Shireman where the cellular incorporation of B(a)P associated "in vitro" with LDL was measured using LDL receptor positive and negative cells and no differences in incorporation were noted between the two cell lines[8]. Taken together the available data may tentatively be interpreted as supporting the concept that VLDL (and probably chylomicron, too) may be good candidates for the major plasma transporters of apolar exogenous compounds. This is also in agreement with the early work by Kotin et al.[9] which suggests that a significant part of B(a)P turnover in the rat may be linked to VLDL and chylomicron turnover.

ACKNOWLEDGEMENTS

The continued interest of Dr. M. Hofnung throughout this study is gratefully acknowledged. Our appreciation is also extended to Dr. N. Weill for her invaluable assistance and suggestions. One of us (J.R.) was supported by a Calouste Gulbenkian Foundation Fellowship.

REFFERENCES

1. E.P. BENDITT and J.M. BENDITT. "Evidence for monoclonal origin of human atherosclerotic plaques". Proc. Natl. Acad. Sci. USA. 70; 1753, 1973.
2. T.A. PEARSON, A. WANG, K. SOLEZ and R.H. HEPTINSTALL. "Clonal characteristics of fibrous plaques and fatty streaks from human aortas", Am. J. Pathol. 81:379, 1975.
3. R.E. ALBERT, M. VANDERLAAN, F.J. BURNS and M. NISHIZUMI.

"Effect of carcinogens on chicken atherosclerosis". Cancer Res. 37: 2232, 1977.

4. H.P. SHU and A.V. NICHOLS. "Benzo (a) pyrene uptake by human plasma lipoproteins <u>in vitro</u>". Cancer Res. 39: 1224, 1979.
5. B.N. AMES, J. McCANN and E. YAMASAKI. "Mthods for detecting carcinogens and mutagens with the Salmonella/mammalian microsome test". Mutation Res. 31: 347, 1975.
6. F.W. ROBERTSON, A.M. CUMMING, A.S. DOUGLAS, E.B. SMITH and A.C.F. KENMURE. "Coronary heart disease in North-East Scotland. A study of genetic and environmental variations in serum lipoproteins and other variables". Scott. Med. J. 25:212, 1980.
7. K.G. TAYLOR, T.J. CARTER, A.J. VALENTE, A.D. WRIGHT, J.H. SMITH and K.A. MATHEWS. "Sex differences in the relationships between obesity, alcohol consumption and cigarett/smoking and serum lipid and apolipoprotein concentrations in a normal population". Atherosclerosis, 38:11, 1981.
8. J.F. REMSEN and R.B. SHIREMAN. "Effect of low-density lipoprotein on the incorporation of benzo (a) pyrene by cultured cells". Cancer Res. 41:3179, 1981.
9. P. KOTIN, H.L. FALK and BUSSER. "Distribution, retention and elimination of ^{14}C-3,4-benzpyrene after administration to mice and rats". J. Natl. Cancer Inst. 23:541, 1959.

FATTY ACID COMPOSITION OF FOUR LIPID CLASSES IN PLASMA OF RATS AND CHICKS RECEIVING LINOLEIC AND α-LINOLENIC ACIDS IN THEIR DIETS*

P. Budowski*, Edna Machtomi,** and I. Bartov**

*Faculty of Agriculture, The Hebrew University
Rehovot, Israel
**Agricultural Research Orginization, the Volcani Center
Rehovot, Israel

ABSTRACT

Weanling rats and one-day-old chicks were fed diets containing 10% of either safflower or linseed oil. Blood was obtained at different times and plasma lipids were separated into triglycerides, nonesterified fatty acids, phospholipids and cholesteryl esters. Feeding linseed oil (53.1% 18:3 and 17.3% 18:2) instead of safflower oil (1.3% 18:3 and 70.0% 18:2) caused extensive replacement of arachidonic acid by 20:5ω3 in all lipid fractions in both species. The cholesteryl ester fraction was species-specific in that its polyunsaturated fatty acids were almost exclusively of the C_{18} type in chicks, whereas rats contained both C_{18} and C_{20} polyunsaturated fatty acids. Modification of the fatty acid profiles of plasma lipids by dietary means could be important in thrombocyte function and thrombogenesis, as well as in cholesterol transport and atherogenesis. It is suggested that in studies in humans, consideration of the relative intakes of ω3 and ω6-polyunsaturated fatty acids would be useful, in addition to the generally used P/S ratio.

INTRODUCTION

The role of oils and fats in human nutrition has been intensively investigated (1) and the desirability of a balanced supply of polyunsaturated fatty acids (PUFA) in relation to saturated fatty acids has been generally recognized (2). In this context, attention has as a rule been directed toward linoleic acid, which is by far the most abundant PUFA in western diets and which is known to be essential to many species (3). Less research has been carried out on the nutritional effects of α-linolenic acid, whose

*Dedicated to Professor A. Bondi on the occasion of his 75th birthday.

essentiality to warm-blooded animals has not been well established (4,5).

We recently reported (6) that linseed oil fatty acids caused an accumulation of ω3-eicosapentaenoic acid in phospholipids of liver and plasma of young chicks. This fatty acid has received attention as a possible antithrombotic agent (7). The purpose of the present study was to extend our previous observations to a more detailed examination of the effect of linseed oil on the fatty acid composition of different plasma lipids in young rats and chicks.

ANIMALS AND PROCEDURES

Weanling male Sprague-Dawley rats weighing about 45 g were fed ad lib. a diet composed of (%): casein (vitamin test) 25, sucrose 55, oil (see below) 10, cellulose 4, choline chloride (22% in casein) 1, vitamins in casein (3) 1, and AIN-76™ mineral mixture (8) 4. The oils used were safflower and linseed oils having the composition shown in Table 1. Groups of four rats were taken at random at the beginning of the experiment and after 17, 27 and 51 days. The animals were fasted for three hours, anaesthesized with ether and bled through the caval vein into test tubes containing sodium oxalate. Plasma was then obtained by centrifugation.

One-day-old male White Rock chicks were kept in thermostatically heated battery brooders and fed ad lib. a diet based on sorghum grain and defatted soybean meal supplemented with 10% safflower or linseed oil. Details on the diet were given previously (9). Blood samples were withdrawn by heart puncture with a heparinized syringe from groups of four chicks before the beginning of the experiment and after four and eight weeks.

Table 1. Fatty acid composition of dietary oils*

Fatty acid	Safflower oil	Linseed oil
Palmitic	8.1	7.0
Stearic	0.4	3.6
Oleic	20.2	19.0
Linoleic	70.0	17.3
α-Linolenic	1.3	53.1

% Of total fatty acids

One or two ml plasma were extracted with chloroform-methanol 2:1 (v/v) and the lipids were submitted to thin-layer chromatography (10). Absorbent areas containing triglycerides (TG), non-esterified fatty acids (NEFA), phospholipids (PL) and cholesteryl esters (CE) were scraped into vials containing 4 ml sulfuric acid in methanol (4% v/v). The vials were sealed with screw caps and heated for one hour in a water bath kept at 80°. After addition of 4 ml water, the methyl esters were extracted with light petrol ether and separated by gas-liquid chromatography (11). Methyl esters obtained from the CE fraction were freed of cholesterol by TLC before GLC. Peaks were identified by comparison with reference standards and by their relative retention times. Peak areas were measured by electronic integration and results are presented as area percentage.

RESULTS

Rat experiment.- The plasma lipids of weanling rats were particularly rich in linoleic and arachidonic acids. The respective values were (%, means±SE, n=4): TG, 38.6±2.1 and 17.4±1.5; NEFA, 33.1±1.1 and 6.0±0.5; and CE, 21.7±0.9 and 35.0±1.2. Rapid changes were induced in the fatty acid composition of plasma lipids as the results of dietary treatments. For the sake of brevity, only the time course for plasma PL is shown (fig. 1). Most of the compositional changes occurred within 27 days after the beginning of the experiment. Safflower oil caused an increase in linoleic acid, but there was no significant change in arachidonic acid . In the linseed oil group 20:5ω3 and 18:3ω3 appeared in the plasma PL, arachidonic acid was greatly depressed and, interestingly, linoleic acid was slightly increased. The PUFA contents of the different plasma lipids at the end of the experiment are presented in fig. 2. Feeding of linseed oil instead of safflower oil caused replacement of arachidonic acid by 20:5ω3 in all fractions. The 20:5ω3 level was especially high in the CE fraction of linseed oil-fed rats where it reached 32.4%. The TG and NEFA fractions tended to reflect the PUFA contents of the dietary oils. The linoleic acid content of the PL and CE fractions, on the other hand, was but little affected by the dietary oils, the effect being without statistical significance. C_{22} PUFA in excess of 1% were found only in the PL fraction of the linseed oil groups, but their total concentration was less than that of 20:5ω3. With these diets the rats reached an average weight of 227 g after 51 days.

Chick experiment.- Chicks receiving the experimental diets grew rapidly and reached an average weight of 2.520 kg at eight weeks of age. The PUFA profiles of the various lipids at the end of the experiment are presented in fig. 3. The results are similar to those found in rats, with the notable exception of the CE fraction which contained virtually only C_{18} PUFA. Comparison of the effect of linseed *vs.* safflower oil reveals that the former caused

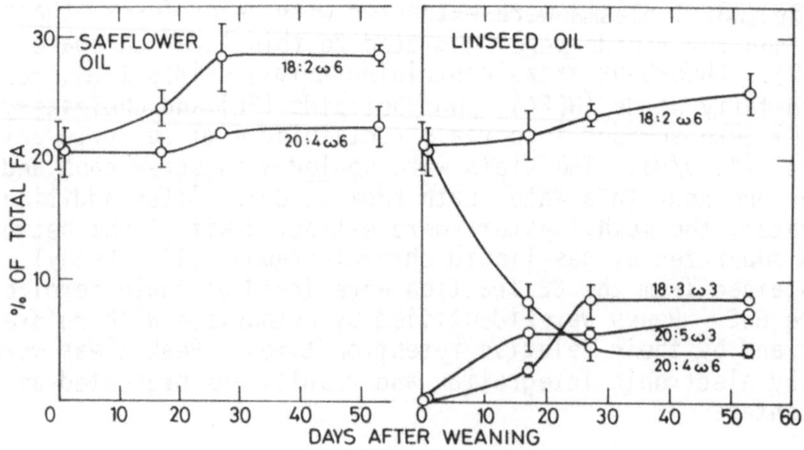

Figure 1. Time course of changes in PUFA levels of rat plasma phospholipids resulting from a dietary supply of 10% of either safflower or linseed oil. Each circle represents the mean of four rats. Standard errors are shown where they do not fall within the circles.

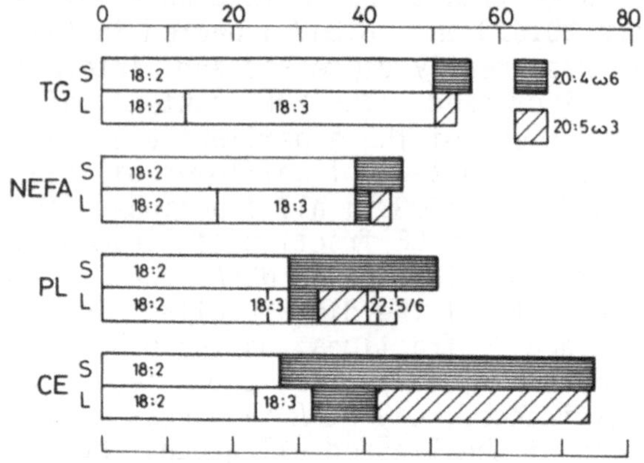

Figure 2. Polyunsaturated fatty acid profiles of triglycerides (TG), nonesterified fatty acids (NEFA). phospholipids (PL) and cholesteryl esters (CE) in plasma of rats receiving 10% of either safflower (S) or linseed oil (L) during 51 days. Minor components (less than 1%) have been omitted. Differences due to dietary oils are not significant for 18:2 in PL and CE; significant (P<0.05) for 20:4ω6 in NEFA; and highly significant (P<0.01) in all other cases.

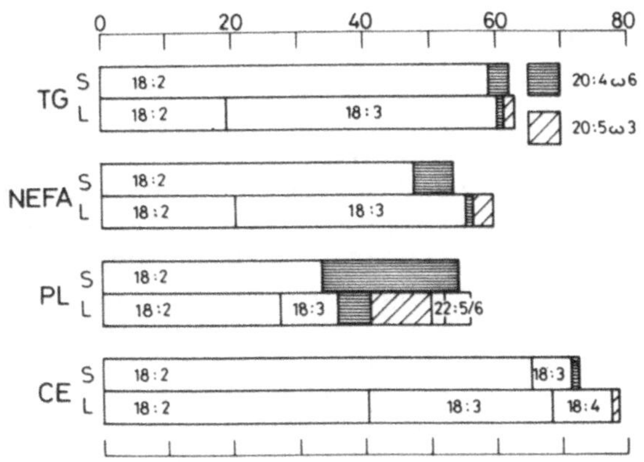

Figure 3. Polyunsaturated fatty acid profiles of triglycerides (TG), nonesterified fatty acids (NEFA), phospholipids (PL) and cholesteryl esters (CE) in plasma of chicks receiving 10% of either safflower (S) or linseed oil (L) during eight weeks. Minor components (less than 1% have been omitted. Differences due to dietary oils are not significant for 20:4ω6 and 20:5ω3 in CE and for 18:2 in PL; and highly significant ($P < 0.01$) in all other cases.

a lower 18:2 level in this fraction, as well as a higher 18:3 concentration and the appearance of a fatty acid with an equivalent chain length of 18:4ω3. Similar effects of the dietary oils on plasma lipid PUFA of chicks were already observed at 4 weeks of age (not shown).

DISCUSSION

The present results have shown that profound differences in the PUFA makeup of plasma lipids can be achieved in young rats and chicks by either dietary safflower oil, which is rich in linoleic acid but contains only about 1% α-linoleic acid, or linseed oil in which the α-linoleic acid content is three times that of linoleic acid. One of the more striking results of a dietary supply of linseed oil is the emergence of 20:5ω3 as the major long-chain PUFA in plasma lipids, especially in the rat CE fraction. This fatty acid largely replaced the arachidonic acid found in the plasma lipids of animals receiving safflower oil.

It has recently been suggested that the presence of 20:5ω3 instead of 20:4ω6 in plasma lipids is partly responsible for the low incidence of cardiovascular disease, reduced platelet aggregability and elevated bleeding times of Greenland Eskimos (12,13), whose diet consists mainly of marine foods poor in ω6-PUFA but containing 20:5ω3 (13). Indeed, ingestion of fish or fish oils by volunteers lead to a rapid increase in the 20:5ω3/20:4ω6 ratio (14,15), while platelet aggregability by low doses of collagen and thromboxane B_2 formation by platelets were decreased (16). Similarly, rats receiving dietary supplements of methyl α-linolenate exhibited decreased platelet TXB_2 production (17). We have recently suggested (6) that the protective effect of linseed oil fatty acids against chick nutritional encephalomalacia may be the result of an antithrombotic effect caused by a rise in 20:5ω3 and a fall in 20:4ω6 in plasma.

The mean ratios of 20:5ω3 to 20:4ω6 in plasma lipids, calculated from the present data, are presented in Table 2 for rats and chicks fed linseed oil. The values are greater than one in all cases. In animals fed safflower oil, the corresponding values were less than 0.03, except for chick CE which had a ratio of 0.25. The plasma PL fraction of Greenland Eskimos has a ratio of about 9, compared to less than 0.03 in Danes (12).

Why do dietary supplements of α-linolenate or linseed oil lead to the formation of 20:5ω3 in preference to 22:6ω3 which predominates among the long-chain ω3 derivatives of plasma and tissues in normally fed animals (18)? Mohrhauer and Holman (19) found that when rats were fed a purified diet supplemented with increasing amounts of ethyl ω-linolenate, 22:6ω3 was the principal long-chain ω3 derivative in liver as long as the dietary 18:3ω3 supplied less than 2% of the calories. When this intake level was exceeded, hepatic 22:6ω3 remained nearly constant while 20:5ω3 continued to

Table 2. Mean ratios of 20:5ω6 to 20:4ω6 in plasma lipids of linseed oil-fed rats and chicks at the end of the experiment*

	Rats	Chicks
Triglycerides	7.8	1.9
Nonesterified fatty acids	1.8	2.4
Phospholipids	1.7	1.9
Cholesteryl esters	3.3	14

* Corresponding values of safflower oil-fed animals were less than 0.03, except for the CE fraction of chicks which had a value of 0.25

increase and became the predominant long-chain PUFA. Under these conditions, the conversion of 20:5ω3 to 22:5ω3 appears to be rate-limiting (20). Stock diets usually provide low concentrations of ω3-PUFA, whereas in the present experiments the dietary intake of 18:3ω3 was approximately 15 Cal%, thus causing an accumulation of 20:5ω3.

The low levels of 22:6ω3 in plasma lipids of rats and chicks receiving linseed oil contrasts with the plasma PUFA profiles found in animals and humans ingesting fish or fish oils, which provide both 20:5ω3 and 22:6ω3 and which result in the accumulation of 22:6ω3, often in excess of 20:5ω3, in PL and TG of liver and plasma (6,14,15,21). Eskimos keep their plasma levels of 22:6ω3 low (12), possibly because of an adaptive mechanism of retroconversion of 22:6ω3 to 20:5ω3, as demonstrated in rats (22). Hornstra et al.(23) have called attention to potentially detrimental effects resulting from the ingestion of excessive amounts of 22:6ω3 and long-chain monoenes contained in fish oils.

α-Linolenic acid competitively inhibits the Δ6-desaturation of linoleic acid, thus decreasing the formation of arachidonic acid (24,25). This effect is apparent in Figs. 1,2 and 3. A concomitant increase in the linoleic acid levels, although only hinted at by the present results, has been clearly demonstrated in rats receiving a constant amount of linoleic acid with increasing supplements of methyl α-linolenate (17). This "sparing effect" of α-linolenic acid on linoleic acid in liver and plasma deserves attention, as it may affect cholesterol transport and atherogenesis.

The present data have emphasized the profound effects exerted by dietary α-linolenic acid on PUFA metabolism and PUFA profiles of plasma lipids. Because of their possible significance in health and disease (1,2,26), these effects should also be investigated in humans for which little information is available (27). Moreover, the ratio of PUFA to saturated fatty acids (P/S ratio) commonly used in experimental and epidemiological nutrition studies fails to take into account the differences in nutritional-metabolic effects caused by dietary linoleic and α-linolenic acids. Such differences ought to be acknowledged by also considering the ratio of linoleic to α-linolenic acid or the ω6/ω3 ratio.

REFERENCES

1. Vergroesen, A.J. (1975) " The Role of Fats in Human Nutrition", Academic Press, London.
2. FAO/WHO (1980) "Dietary Fats and Oils in Human Nutrition", Food and Agriculture Organization of the United Nations, Rome.
3. Holman, R.T. (1968) in " Progress in the Chemistry of Fats and Other Lipids", vol. 9, edited by R.T. Holman, Pergamon, Oxford, pp. 274-348.
4. Lemarchal, P. (1978) REV. franc. Corps Gras 25, 303-308.
5. Tinoco, J., Babcock, R., Hincenbergs, I., Medwadowski, B., Miljanich, P. and Williams, M.A. (1979) Lipids 14, 166-173
6. Budowski, P., Hawkey, C.M. and Crawford, M.A. (1980) Ann. Nutr. Alim. 34, 389-400.
7. Dyerberg, J., Bang, H.O., Moncada, S. and Vane, J. R. (1978) Lancet ii, 117-119.
8. AIN Report (1977) J.Nutr. 107 , 1340-1348.
9. Bartov, I., Lipstein, B. and Bornstein, S. (1974) Poultry Sci. 53, 115-124.
10. Sklan, D., Hurwitz, S., Budowski, P. and Ascarelli, I.(1975) J. Nutr. 105, 57-63.
11. Sklan, D., Volcani, R. and Budowski, P. (1971) J. Dairy Sci. 54 , 515-519.
12. Dyerberg, J., Bang, H. O. and Hjorne, N. (1975) Am. J. Clin. Nutr. 28 , 958-966.
13. Dyerberg, J. and Bang, H.O. (1979) Haemostasis 8, 227-233.
14. Von Lossoncy, T.O., Ruiter, A., Bronsgeest-Shoute, H.C., van Gent, C.M. and Hermus, R.J.J. (1978) Am. J. Clin. Nutr. 31, 1340-1346.
15. Van Gent, C.M., Luten, J. B., Bronsgeest, H.C. and Ruiter, A. (1979) Lancet ii, 1749-1750.
16. Siess, W., Scherer, B., Bohling, B, Roth, P., Kurzmann, I. and Weber, P.C. (1980) Lancet i, 441-444.
17. Hwang, D.H. and Carroll, A.E. (1980) Am. J. Clin.Nutr. 33, 590-597.
18. White, D.A. (1973) in " Form and Function of Phospholipids", 2nd ed., edited by G.B. Ansell, R.M.C. Dawson and J.N. Hawthorne, Elsevier, Amsterdam, pp. 441-482.
19. Mohrhauer, H. and Holman, R.T. (1963). J. Lipid Res. 4, 151-159.
20. Poovaiah, P. B., Tinoco, J. and Lyman, R.L. (1976) Lipids 11, 194-202.

21. Witting, L. A., Harvey, C.C., Century, B. and Horwitt, M. K. (1961) J. Lipid Res. 2, 412-418.
22. Schlenk, H., Sand, D. M. and Gellerman, J.L. (1969) Biochim.Biophys. Acta 187, 201-207.
23. Hornstra, G., Haddeman, E. and ten Hoor, F. (1979) Lancet ii, 1080.
24. Mohrhauer, H. and Holman, R.T. (1963) J.Nutr. 81, 67-74.
25. Brenner, R.R. and Peluffo, R.O. (1966) J. Biol. Chem. 241, 5213-5219.
26. Mead, J. F. and Fulco, A. J. (1976) " The Unsaturated and Polyunsaturated Fatty Acids in Health and Disease ", Chas. C. Thomas, Springfield, Illinois.
27. Dyerberg, J., Bang, H.O. and Aagaard, O. (1980) Lancet i, 199.

EFFECTS OF ALCOHOL ON HIGH DENSITY LIPOPROTEINS (HDL)

Mª da Graca Paiva de Carvalho, Pedro O. Rodrigues,
Mª Fernada de Mesquita, João Magalhães, and M.J. Halpern

Dept. of Biochemistry and Center for Lipid Research (INIC)
Faculty of Medical Sciences UNL
Campo de Santana 130, P-1100 - Lisboa - Portugal

INTRODUCTION

The high density lipoproteins (HDL) have been the focus of much interest since their plasma concentrations, commonly measured as HDL-cholesterol, were shown to be inversely correlated with the risk of coronary heart disease (CHD) (1-5).

Among several factors, it is reported the positive action of the alcohol (6) in increasing the plasma HDL cholesterol. The goal of this paper it's to study the effects of 4 dl of table wine, in normal persons, type IV and type IIb hiperlipemia.

PATIENTS AND METHODS

A - Selection of Patients

The blood samples were obtained by venopuncture of the forearm in healthy persons, Fredrickson's type IV and type IIb patients and type IV normo triglicerydemic.

The subjects were submeted to 4 dl of table wine (red wine of 12º ± 38,4 g of alcohol) ingestion and blood was drawned at 0, 30, 60, 90 and 120 minutes.

B - Methods

Alcohol was determined by enzimatic method (7) and HDL cholesterol by precipitation with phosphotungstic acid (8,9) followed by CHOD-PAP method (10) for total cholesterol.

RESULTS

In the normal persons (fig.1) after the wine load, the fast HDL-cholesterol of 32 mg/dl increased to 47 mg/dl at 90' which was coincident with the maximum alcohol peak.

The response of Fredrickson's type IV is represented in fig. 2.

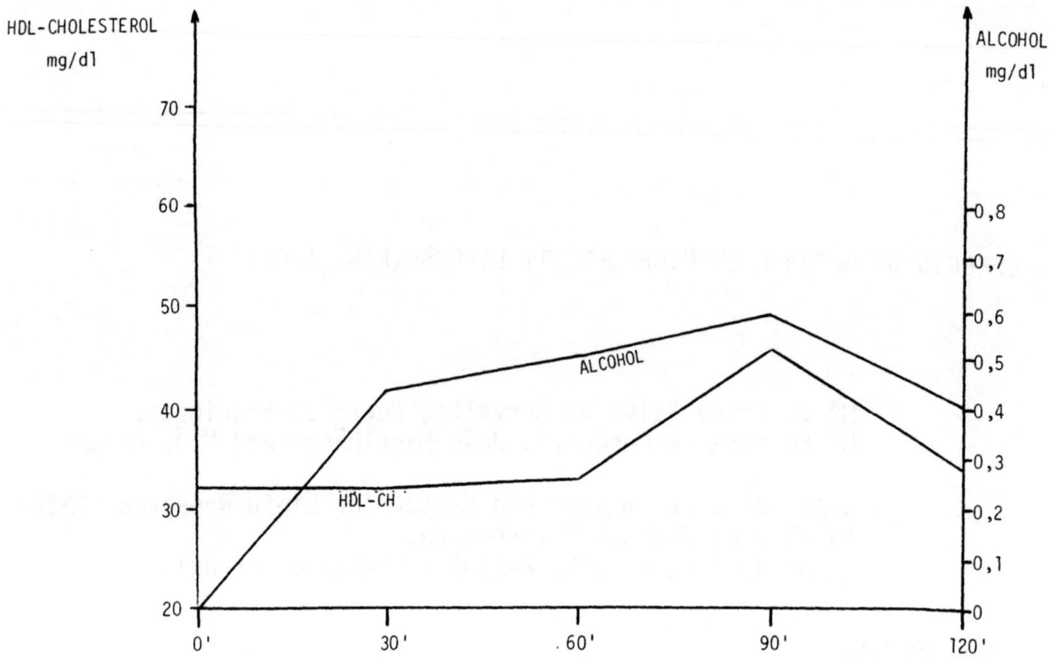

Figure 1. Serum Alcohol and HDL cholesterol response to 4 dl load of wine in healthy persons.

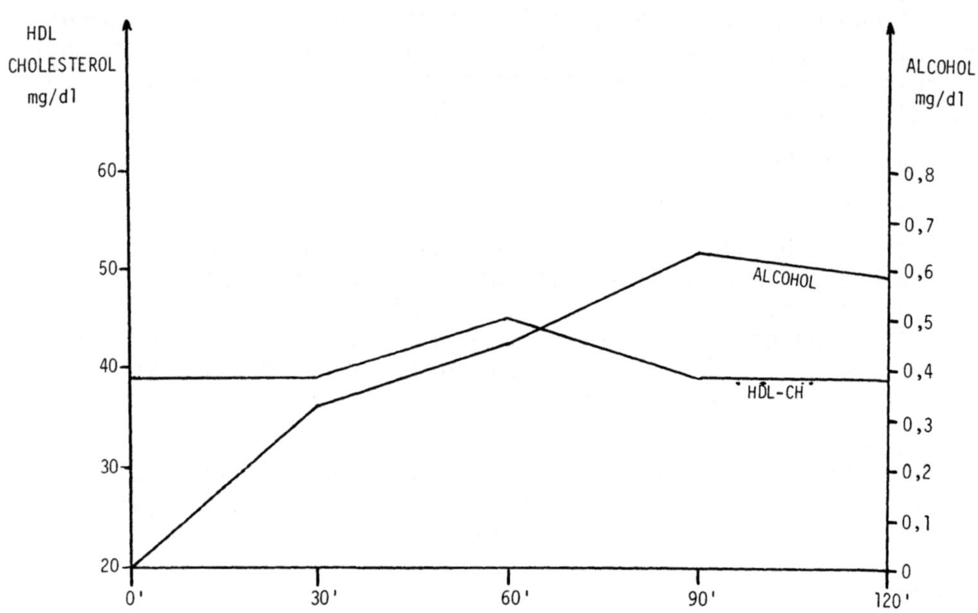

Figure 2. Serum alcohol and HDL cholesterol response to 4 dl load of wine in Fredrickson's type IV.

The fasting HDL-cholesterol of 39 mg/dl increased and reached its peak at 60' with 45 mg/dl, while the alcohol peak was at 90'.

In the type IV normotriglicerydemic (fig. 3) the HDL cholesterol of 47 mg/dl increased to 50 mg/dl, having the maximum peak for HDL-cholesterol and alcohol both at 90'.

The last case studied a type IIb (fig 4), after the wine load, the HDL-cholesterol decreased to reach the minemus value at 90' while the alcohol reached its maximus peak. (Fasting HDL--cholesterol: 33 mg/dl and 90' HDL-cholesterol: 30 mg/dl.
Four weeks later the patient's electrocardiogram showed an inferior infarction, and coronary test detected (fig 5).

DISCUSSION

In our small study it's confirmed the positive inducing factor of the alcohol on HDL-cholesterol to wich until now there is no explanation, as evidentiated by the last case that we have reported.

How does alcohol works? that's the reason of our group works for searching an answer to this complexe question.

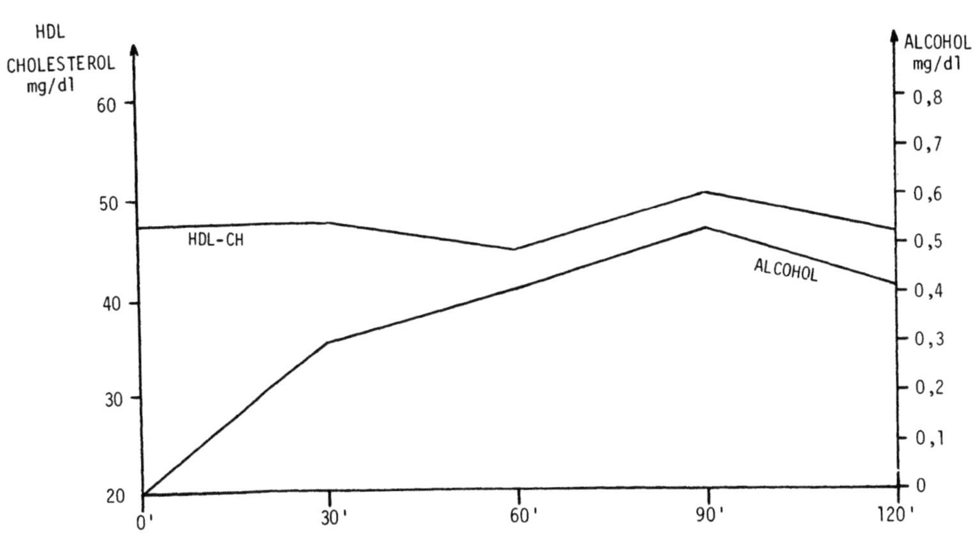

Figure 3. Serum alcohol and HDL-cholesterol response to 4 dl load of wine in type IV normotriglicerydemic.

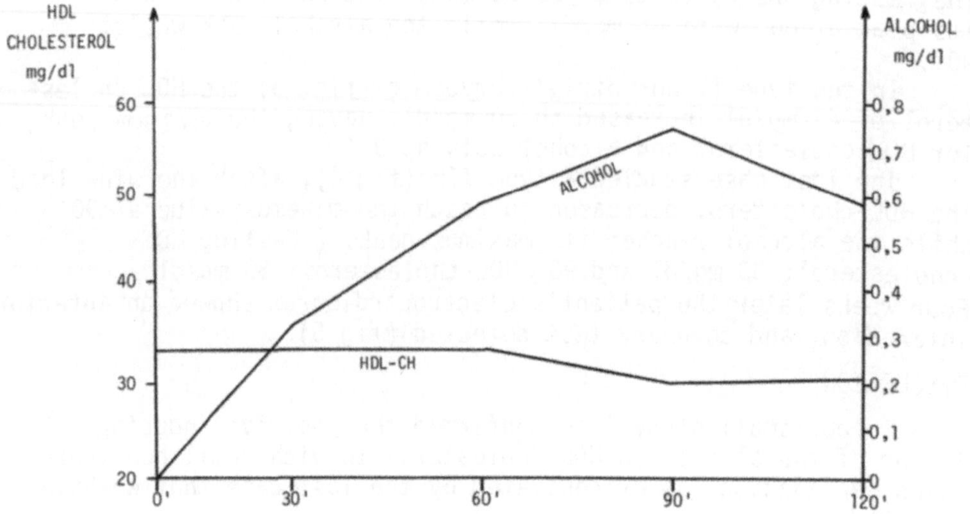

Figure 4. Serum alcohol and HDL-cholesterol response to 4 dl load of wine in Fredrickson's type IIb

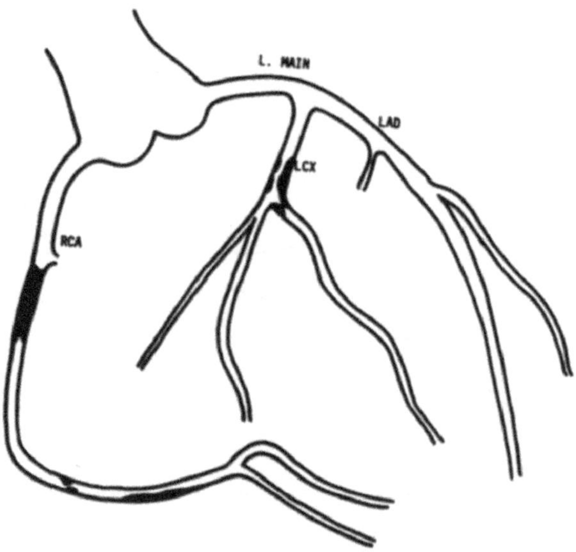

Figure 5. Coronary test.

REFERENCES

1. Rhoads, G.G., Gulbrandsen, C.L. and Kragan, A: Serum lipoproteins and coronary heart disease in population study of Hawai Japonese men. N.Engl. J. Med 294:293-98 1976
2. Gordon T, Castelli W.P. Hyortland, M.C. et al.:High density lipoprotein as a protective factor against coronary heart disease. Am.J.Med 62:707-14,1977.
3. Streger D. Steiner, G. and Kwiterowich, P.O., Jr: Plasma high density lipoproteins and ischemic heart disease. Ann. Intern. Med 89:871-80, 1978.
4. Miller, M.E., Thelle, D.S. Forde, O.H. and al: The Tromso heart study. High density lipoprotein and coronary heart disease: a prospective case control study. Lancet 1:965-67,1977.
5. Kannel, W.B., Castelli, W.P. and Gordon, T: Cholesterol in the prediction of atherosclerotic disease, Ann. Intern. Med. 90:85-91, 1979.
6. Yano K. Rhoads G.G. and Kagana. Coffee, alcohol and risk of coronary heart disease among Japonese men living in Hawai. New Engl. J.Med (1977) 297, 405-409.
7. Bencher, Th and H. Redetzki Klin. Wschr. 29 (1951) 615.
8. Lopez Virella M.F. and al. - Clin. Res. 23:882, 1977
9. Burnstein M. and al.: Lipid Res. 21:583, 1970
10. Roschlau P. and al. : Ann. Clin. Bichem. 6:24, 1975.

DIAGNOSIS AND TREATMENT OF HYPERLIPIDEMIAS

Jean Davignon

The Dept. of Lipid Metabolism and Atherosclerosis Research, Clinical Research Institute of Montreal
The Section of Vascular Medicine, Hôtel-Dieu Hospital Montreal - The Dept. of Medicine - Faculty of Medicine University of Montreal

INTRODUCTION

The central role of plasma cholesterol in atherogenesis has been amply demonstrated (1). It is also well established that hyperlipidemia is an important risk factor for coronary heart disease (CHD). In addition there is considerable evidence in experimental animals (1,2) and in man (1,3-5) that lowering of plasma lipids will reduce the cardiovascular complications of atherosclerosis and promote regression of the lesions. The physician is thus justified to make an all out effort to reduce elevated plasma lipid levels in the hope of preventing, arresting, or reverting the atherosclerotic process. This is especially important when a diagnosis of hereditary hyperlipidemia is established since in many instances, as in familial hypercholesterolemia (FH), the affected individuals may be at risk of developing cardiovascular complications at an early age. In other situations (lipoprotein lipase deficiency for instance), the physician might want to lower the plasma lipids to prevent an acute attack of pancreatitis or to bring about a regression of unsightly xanthomatous skin lesions. To be effective, any approach to the treatment of familial hyperlipidemias must be rational and systematic. I would like to review here the major steps to be taken in such a therapeutic program (6,7).

First step: an accurate diagnosis

An accurate diagnosis is mandatory if an appropriate treatment is to be instituted. Indeed the nature and the extent of the treatment will depend to a large degree on the type and severity of the plasma lipid transport disorder. This implies a complete physical examination coupled with appropriate laboratory tests to exclude secondary causes of hyperlipidemia and aggravating factors, get more insight into the plasma lipoprotein abnormality and assess its potential impact in terms of cardiovascular risk (Table 1).

Many of the primary hyperlipidemias are not immediately obvious and their detection may be entirely fortuitous. It is thus most useful to look for the clinical clues listed in Table 2. The family history should be carefully scrutinized for the presence of coronary heart disease (CHD), peripheral atherosclerosis (PVD) or cerebrovascular events (CVE). An effort should be made to detect the existence of premature atherosclerosis in both the patient and his family. The physician

Table 1. CLINICAL AND LABORATORY WORK-UP FOR THE DIAGNOSIS OF FAMILIAL HYPERLIPIDEMIAS AND DYSLIPIDEMIAS.

<u>Clinical</u>:

Complete history and physical examination
Family history
Dietary habits and medication

<u>Standard laboratory tests and X-rays</u>:

Cholesterol, triglycerides, uric acid
Fasting blood sugar
Liver and renal function tests
Urinalysis
Thyroid function tests including TSH
Glycemia 2h after a 100 g glucose load
ECG (stress test if indicated)
Flat plate of abdomen (AP and L)

<u>Specialized tests</u>:

Determination of lipoprotein cholesterol
(LDL-C, HDL-C, LDL/HDL, VLDL-C/TG, β-VLDL, Lp(a), slow preβ)
Plasma lipoprotein lipase activity
Polyacrylamide gel electrophoresis of apo C
Isoelectric focusing of apolipoprotein E of VLDL
Apolipoprotein B of plasma LDL and VLDL
LCAT assay
Plasma β-sitosterol and cholestanol

must pay close attention to the eyes (fundus, cornea, eyelid), the skin of elbows, knees, palms of the hands and buttocks and the tendons where signs of lipid infiltration may be detected. Arcus corneae in young adults usually reflect the presence of hyperlipidemia with the exception of certain black populations (8) and are considered by some as an independent risk factor (9). Sometimes they are incomplete and appear as a discrete whitish crescent which may be masked by the eyelid and missed if the physician does not take time to lift the eyelid and look closely when shining a light laterally on the cornea. About half of individuals with typical xanthelasma, even young adults, may have normal plasma lipid levels. We have recently found (10) that normolipidemic patients with xanthelasma have circulating lipopoteins of abnormal composition: elevated apolipoprotein B in LDL, and/or apolipoprotein E_2/E_3 ratio in VLDL compatible with the heterozygous state for familial dysbetalipoproteinemia (apo E-ND phenotype). This condition should not be dismissed as benign since an increased incidence of CHD has been reported in normolipidemic individuals harboring xanthelasma of the eyelids (11). Care must be taken not to confuse xanthelasmas with syringomas. The latter, benign tumors of the sweat glands are more brownish, spotty and less confluent. Ophthalmoscopic

Table 2. USEFUL CLUES FOR THE DIAGNOSIS OF PRIMARY HYPERLIPIDEMIAS.

1- Family history of premature atherosclerosis or of a high frequency of cardiovascular complications (CHD, PVD, CVE)
2- Presence of the manifestations of atherosclerosis in the patient
3- Arcus corneae
4- Xanthelasma
5- Lipemia retinalis
6- Xanthoma eruptivum
7- Attacks of abdominal pain in childhood
8- Pigmentation of palmar creases, brown or orange (xanthochromia striata palmaris)
9- Planar xanthomas along palmar creases (orange or yellowish)
10- Punctate xanthomas of hands or fingers
11- Tuberoeruptive xanthomas
12- Tuberous xanthomas
13- Xanthoma tendinosum (fingers, heels)
14- Plantar xanthomas
15- Extensive planar xanthomas, intertriginous xanthomas, thick xanthomas of creases
16- Arterial bruit
17- Turbid, lactescent or creamy plasma

examination of the fundi should be part of the examination since it may reveal a typical lipemia retinalis which is complete when plasma triglycerides exceed 5000 mg/dl. Eruptive xanthomas in a young person might indicate the presence of a severe hypertriglyceridemia (2000 mg/dl or more) most often related to a hyperchylomicronemia. When associated with a history of abdominal pain in childhood, the diagnosis of familial lipoprotein lipase deficiency (type I) should be suspected.

A careful examination of the hands may reveal lesions which are specific for familial dysbetalipoproteinemia (type III). Planar xanthomas along the creases of the palms (Figure 1), brown or orange pigmentation of the creases (xanthochromia striata palmaris) and punctate xanthomas at the cubital edge of the hand or at the tip of the fingers, although often discrete, are quite typical of the disease (12). Tubero-eruptive xanthomas of the elbows, or eruptive xanthomas of the buttocks tend to have a brownish-orange color in type III. Tuberous xanthomas (elbows, knees or heels) although often seen in this disease are not specific and may be found also in familial hypercholesterolemia (FH, type II) and hypertriglyceridemia (type IV). Tendon xanthomas may be found in familial type III (about 25% of cases) and in familial hyperprebetalipoproteinemia (type IV) but they are usually the landmark of familial hypercholesterolemia (12,13,14). They may be present at the Achilles' tendons and/or along the extensor tendons of the fingers. If the physician however does not take time to palpate these structures carefully, he will miss the early nodular lesions. Occasionally, patients with FH are referred by orthopedic surgeons because of a lump under the foot rendering walking uncomfortable (xanthoma of the plantar aponeurosis). The patient with homozygous familial hypercholesterolemia may be readily detected because of the severity and early expression of his xanthomatous lesions (Figure 2). For instance, tuberous xanthomas of the gluteal fold may be present at birth and intertriginous lesions of the dorsal aspect of the hands (15) and thick planar xanthomas of the popliteal region are quite typical. In this condition xanthomas may be found along the palmar creases but, in contrast to those of type III, they are raised and quite obvious.

One must not forget to take time to examine a sample of plasma left standing over a 24-hour period to look for creaming (hyperchylomicronemia) or diffuse turbidity (endogenous VLDL particles), remembering that the turbidity of plasma is imparted by the triglyceride-rich particles and begin to be detectable when triglyceride levels exceed 300 mg/dl.

Looking for such clinical clues will be invaluable for detecting the presence of hyperlipidemia, assessing its severity, improving the diagnostic accuracy and guiding the treatment.

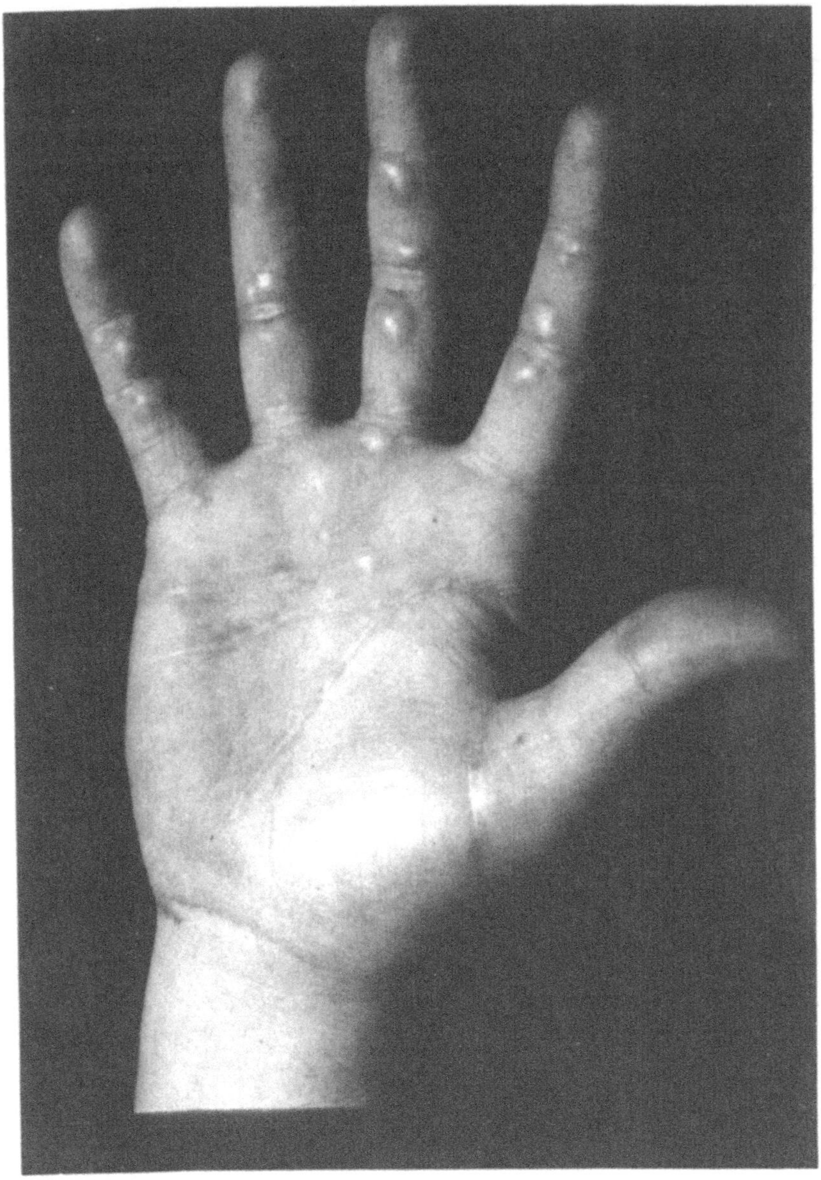

Fig. 1. Xanthomas of the palmar creases in a 26 year old patient with familial dysbetalipoproteinemia (type III). Those on the finger creases tended to fissurate and bleed when he worked with his hands. Usually these nearly pathognomonic xanthomas are much more discrete. They may be yellowish or orange.

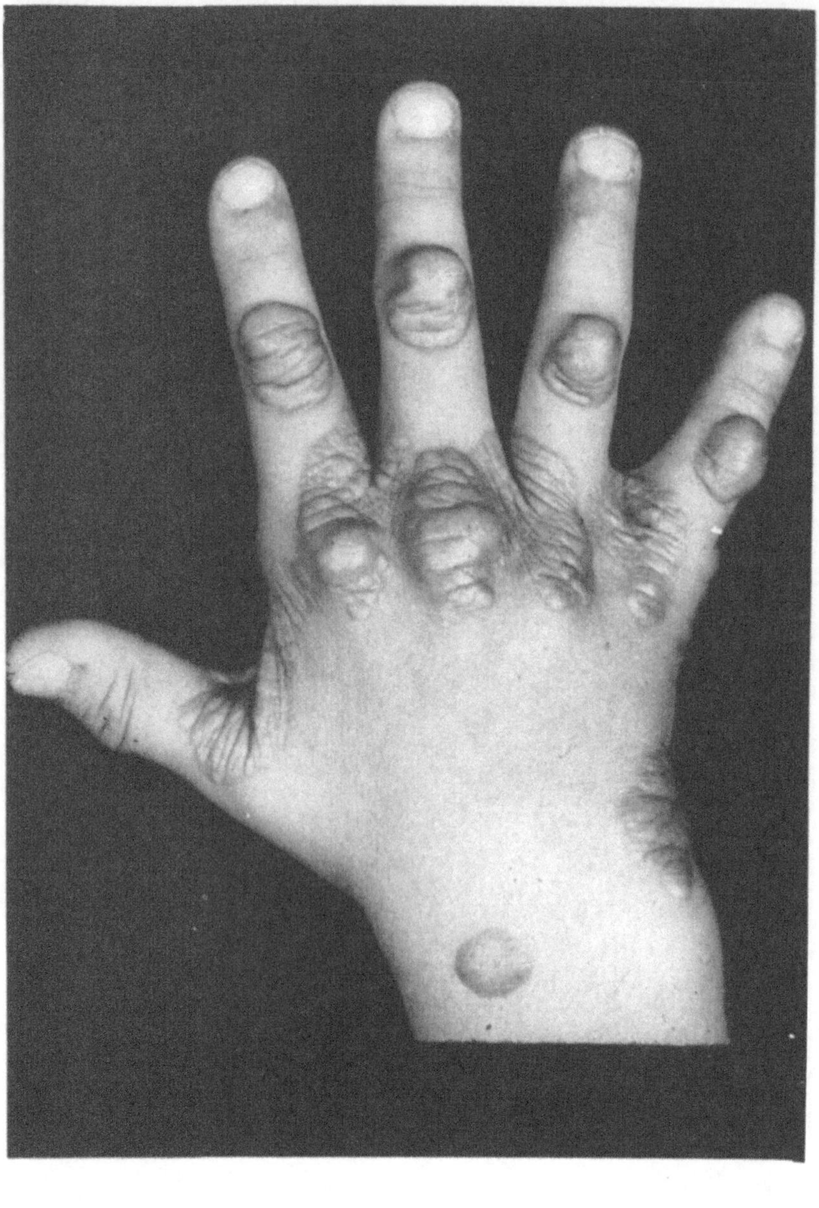

Fig. 2. Tuberous and planar xanthomas of the dorsum of the hands in an 8 year old boy affected with homozygous familial hypercholesterolemia and premature atherosclerosis. The intertriginous lesions are often seen in this disease.

In the process of making a diagnosis, the physician should have in mind a simple and useful classification of hyperlipidemias (Table 3). It should include well defined disease entities and be modified if needed as a function of new advances in knowledge. The lipoprotein phenotyping system (16) based on electrophoresis combined with measurements of cholesterol in the various lipoprotein fractions (Table 1) still provides valuable information on the plasma lipoprotein abnormality in the fasting state. The phenotype however must not be equated with a disease entity. The physician should be aware in particular of some of the difficulties inherent to the interpretation of the lipoprotein phenotype (presence of a sinking prebeta-lipoprotein, of a slow migrating prebeta-lipoprotein, influence of various factors which may induce changes from time to time in the same individual, the notion of multiple phenotype disease, limited significance of the broad-beta band, etc...) (6). Most practicing physicians are familiar with the major plasma lipid transport disorders (14) listed in Table 3 especially familial hyperchylomicronemia, familial hypertriglyceridemia, familial mixed hyperlipidemia and familial hypercholesterolemia. Over the past few years new disease entities have evolved such as familial combined hyperlipidemia, the diagnosis of familial dysbetalipoproteinemia has been refined and more emphasis has been put on the protein moiety of the lipoproteins, the apolipoproteins.

The diagnosis of familial combined hyperlipidemia (FCH) should be based essentially on a family survey which will demonstrate the presence of multiple lipoprotein electrophoretic phenotypes (IIb, IIa, IV and even V) in the kindred (14,17). Otherwise the diagnosis is one of exclusion (Table 4). Usually the increases in plasma cholesterol and triglycerides are rather moderate as compared to the levels found in FH. It is inherited as a dominant monogenic trait with incomplete penetrance and delayed expression in childhood. Few clinical signs are present and tendon xanthomas have not been reported. CHD indidence is high but tends to occur later in life than in FH (18) and there is no LDL-receptor defect (14).

Some of the specialized laboratory tests listed in Table 1 may be helpful to confirm the diagnosis of familial dysbetalipoproteinemia (type III). Often plasma cholesterol (C) and triglyceride (TG) concentrations are elevated to the same extent with cholesterol levels usually quite high (in the 400 mg/dl range). Combined ultracentrifugation, heparin-manganese precipitation and lipoprotein electrophoresis (20,21) are helpful in demonstrating the β-VLDL which is enriched in cholesterol esters and apolipoprotein E (apo E). These techniques allow for the determination of the VLDL-C/TG ratio which is diagnostic if greater than 0.30 when plasma triglycerides are under 1000 mg/dl (22). The presence of a broad beta band on paper or agarose

Table 3. A SIMPLE AND PRACTICAL CLASSIFICATION OF FAMILIAL (F) HYPERLIPIDEMIAS AND DYSLIPOPROTEINEMIAS.*

A- <u>Preponderant hypertriglyceridemia</u>

1- F. Hyperchylomicronemia, Lipoprotein Lipase Deficiency, "type I" (I)

2- F. Apolipoprotein C-II Deficiency (V, I)

3- F. Hypertriglyceridemia, F. Hyperprebetalipoproteinemia, "type IV" (IV, V)

4- F. Mixed Hypertriglyceridemia, "type V" (V)

5- Latent Hyperlipidemia (N → IV or V)

B- <u>Preponderant hypercholesterolemia and dyslipoproteinemias</u>

6- F. Combined Hyperlipidemia, Multiple Phenotype Disease, (IIb, IIa, IV and V)

7- F. Hyperlipidemia with slow-migrating pre-β band (IIb)

8- F. Dysbetalipoproteinemia, Broad-Beta-Disease, Remnant Disease, "type III" (III, IV, V, IIb)

9- F. Hypercholesterolemia, LDL-Receptor Disease, "type II" (IIa, IIb)

10- Pseudo-Homozygous F. hypercholesterolemia

C- <u>Usually normolipidemic dyslipoproteinemias</u>

11- F. Hyperalphalipoproteinemia (N, ↑ HDL)

12- F. Hyperapobetalipoproteinemia (N, IIa)

13- F. Hyperbetasitosterolemia (N)

14- Cholestanolosis, Cerebrotendinous Xanthomatosis (CTX) (N)

---

* Fourteen disease entities and some of their synonyms are listed shorthand name derived from the classification of Fredrickson et al. is given between brackets, the lipoprotein phenotypes are given in parentheses.

Table 4. DIAGNOSIS OF FAMILIAL COMBINED HYPERLIPIDEMIA (FCH)

1- Few clinical signs, no tendon xanthoma reported

2- Increases in plasma cholesterol and/or triglycerides

3- Plasma lipid levels lower than in familial hypercholesterolemia

4- Various lipoprotein phenotypes in the family (IIb, IIa, IV, V)

5- Coronary Heart Disease (50-60y)

6- Dominant monogenic inheritance with low penetrance (delayed expression in childhood)

7- No LDL-receptor defect

8- Better response to treatment than FH (?)

electrophoresis is not very useful since it is present in only 46% of cases (23). More recently it has been shown that plasma apo E concentrations are elevated (24) and that type III is characterized by a deficiency of one of the isomorphs of apo E separated by isoelectric focusing of delipidated VLDL (25). Demonstration of a deficiency in apo E_3 (apo E_2/E_3 ratio 3.82) is the most widely accepted criterion at present for the diagnosis of type III. Furthermore it is possible with this technique to detect heterozygotes for this recessive disease (apo E_2/E_3 ratio between 0.72 and 2.29). Recently, Breslow and Zannis (26) have used two-dimensional polyacrylamide gel isoelectric focusing and electrophoresis to demonstrate a typical pattern they label "β-IV". It is unfortunate that most of these techniques are not available in hospital clinical laboratories at present.

The knowledge that apolipoprotein C-II, which is mostly contained in plasma VLDL, is a major activator of plasma lipoprotein lipase (27) has led Breckenridge et al. (28,29,30) in Canada to discover a new disease entity, familial apo C-II deficiency. It is characterized by marked hypertriglyceridemia with elevated chylomicrons with or without increased VLDL concentrations (phenotypes I or V), low LDL-C and HDL-C, a high incidence of pancreatitis and an autosomal recessive mode of inheritance. Correction of the hyperlipidemia by intravenous infusion of plasma or of apo C-II distinguishes it from familial lipoprotein lipase deficiency. Familial hyperapobetalipoproteinemia is another recently recognized disease entity related to abnormal apolipoprotein metabolism (31). The high plasma levels of LDL apolipoprotein B (apo B) even in the absence of marked plasma cholesterol elevation appear to be responsible for a clinical picture which resembles that of homozygous familial hypercholesterolemia. This finding is in line with the presumed atherogenicity of apo B reported in CHD patients with normal plasma lipid levels (32). Familial hyperapobetalipoproteinemia is inherited as an autosomal dominant disease and the homozygote also has hyperbetasitosterolemia (31). This is of interest in view of a former report linking familial hyperbetasitosterolemia to the presence of xanthomatous lesions in the absence of plasma cholesterol or triglyceride elevation (33). Finally, hyperalphalipoproteinemia is another familial entity where apolipoproteins A-I and A-II are elevated as well as HDL-cholesterol (19,34). The latter is greater than 70 mg/dl and may reach 160 mg/dl. Plasma cholesterol may be normal or slightly elevated. It should not be treated by a lipid-lowering regimen since this anomaly appears to confer a longer life expectancy, it is thus very important to recognize its presence. Indeed if the family history reveals a large number of relatives living to be 80 or more and a low frequency of CHD, the physician must suspect that hyperalphalipoproteinemia is responsible for the mild elevation of plasma cholesterol in his patient.

A latent hyperlipidemia which may be of significance in terms of CHD incidence has been reported by De Gennes (35). Affected subjects are normolipidemic after a 12-hour fast but present hypertriglyceridemia after an 8-hour fast. Pseudohomozygous hypercholesterolemia (36) is a rather obscure disease where the affected child has the skin manifestations found in homozygous FH but both his parents are normolipidemic. In contrast to homozygous FH however the high plasma cholesterol and LDL-C levels are highly responsive to a cholesterol lowering diet.

Finally cerebrotendinous xanthomatosis (37) is also very rare. It is characterized by xanthomas (tendons, lung, brain), subnormal intelligence (ad dementia), progressive cerebellar ataxia, spinal cord paresis, cataracts and atherosclerosis. It appears to be related to elevated levels of plasma cholestanol (1.3 to 4 mg/dl) with usually normal plasma cholesterol and triglycerides.

While the patient is being investigated and before applying any form of treatment it is very important to obtain enough plasma lipid determinations (3 to 5) to establish a baseline for properly assessing later on, the effectiveness of the plasma lipid-lowering measures. Making a judgment from too few lipid measurements and treating an artefact (laboratory errors) are pitfalls to be avoided. One must also beware of the presence of aggravating factors such as large alcohol intake, use of oral contraceptives, hypothyroidism, diabetes and other secondary causes of hyperlipidemia. Excessive alcohol intake for instance may raise plasma triglycerides considerably in normolipidemic obese individuals and more so in those who already have one form of familial hyperlipidemia. One must remember that primary and secondary causes may coexist in certain subjects both contributing to the hyperlipidemia. This will become evident with the persistance of a residual hyperlipidemia when the secondary cause is taken care of. Failure to be attentive to these details can result in the physician making a wrong diagnosis.

Second step: control other risk factors

It is very disappointing to see physicians investing a large amount of time and effort trying to lower plasma cholesterol or triglyceride levels while completely forgetting about the control of hypertension, cigarette smoking, diabetes, obesity and life style. Not only is it important to do that from the start, but it is also mandatory to follow blood pressure at subsequent visits, since sometimes the onset of hypertension may indicate that a plaque has developed on a renal artery. In the control of hypertension, one must remember that certain drugs such as thiazides (38) and propranolol (39) tend to raise plasma lipid levels. Some β-blockers such as metoprolol (39) do not share the triglyceride-raising and the HDL-cholesterol-lowering effects of propranolol. Obesity should be a primary target since its importance as a risk factor is linked to the increased incidence of associated conditions which are themselves considered to be risk factors for CHD, such as arterial hypertension, hypertriglyceridemia, glucose intolerance, hyperinsulinism, hyperuricemia and sedentarism (1).

Third step: dietary management

Dietary management constitutes the backbone of the treatment of primary hyperlipidemias. This is where the physician is most likely to meet with resistance and lack of compliance (40,41,42). This aspect should be approached positively, taking into account the socio-cultural context of the patient and should be done with the help of a dietitian. Certain golden rules are to be followed. First, one must make sure that the instructions are given in the presence of the member of the family who is responsible for preparing the meals. Second, a diet is not to be prescribed until the dietary habits have been thoroughly investigated, the percentage of calories derived from protein, fats, sugars and alcohol calculated, and the daily ingestion of calories, cholesterol, saturated fats and simple sugars well established. Third the compliance should be monitored at each visit and at each visit the instructions are to be reinforced and the patient encouraged. Fourth, the visits should not be scheduled too far apart, an essential element to insure compliance (the "out of sight-out of mind" rule).

The diet should be tailored to the plasma lipoprotein abnormality (6,7,42). Excess body weight must be given attention first and corrected by reducing calories. Lowering dietary saturated fat and cholesterol, and increasing the intake of polyunsaturated fat (especially linoleic acid) are essential to achieve an adequate control of hypercholesterolemia when it is secondary to an increase in LDL-cholesterol. Intake of simple carbohydrates and alcohol should be restricted when there is an increase in endogenous triglycerides and VLDL-cholesterol. In many instances success may be achieved by dietary control alone especially in obese individuals (41). One must remember that in the non-obese subject an isocaloric dietary restriction in simple sugar will promote a reduction of endogenous triglycerides (41).

Raising the linoleic acid content of the diet might be beneficial in more than one way (42,43). This fatty acid not only lowers both cholesterol and triglycerides but has been shown to have antiatherogenic properties in animals, to inhibit platelet aggregation and exhibit anti-thrombogenic effects. More recently data obtained in the experimental animal and in epidemiologic studies indicate that it has a mild hypotensive effect. Furthermore, it might improve myocardial function and be beneficial to a defective insulin metabolism in obese subjects.

Fourth step: the use of drugs

General guidelines. The decision to use drugs must be made only after completing an adequate dietary trial and demonstrating that it has failed to provide a satisfactory control of plasma lipid levels. The response to the diet may be assessed properly usually within 6 to 12 months if enough plasma lipid determinations (4 or more) have been obtained before and during treatment.

The notion of "satisfactory control" is not absolute and depends a great deal on the clinical setting. Arbitrarily one may be satisfied when levels of cholesterol and triglycerides of less than 240 mg/dl and 150 mg/dl respectively have been attained. On the other hand other major objectives must also be sought such as: ideal body weight in obese subjects, freedom from pancreatitis in severe hypertriglyceridemia, regression of xanthomatous lesions, improvement in the clinical manifestations of atherosclerosis and ultimately demonstration of plaque regression.

General aims are needed but the treatment should be

individualized in each case and the physician must be guided in his therapeutic approach by more than pre-selected plasma lipid cut-offs, he must give particular attention to the following factors:

a) The nature and severity of the hyperlipidemia. Each hereditary hyperlipidemia has its own plasma lipoprotein profile which in turn determines much of the clinical consequences and prognosis of the disease. The treatment should take into account the lipid abnormality and be more aggressive where the likelihood of developing coronary heart disease is greatest. One would not want, for instance, to treat familial hyperalphalipoproteinemia which is associated with a greater longevity (34). Familial hypercholesterolemia, on the other hand, is notoriously lethal and resistant to treatment and most of the time should be treated with a combination of drugs, in addition to the diet, before any worthwhile change in plasma cholesterol is achieved. No drugs are currently effective in familial hyperchylomicronemia (type I) whereas familial dysbetalipoproteinemia is strikingly responsive to clofibrate and its derivatives. This again emphasizes the importance of making an accurate diagnosis before undertaking any form of treatment. The fact that the patient already has CHD or cumulates a number of other risk factors for CHD does worsen the prognosis and should also influence the therapeutic attitude.

b) The familial context. The morbidity and mortality from cardiovascular diseases among the first and second degree relatives of the patient should constitute one of the most important guide in the decision to treat and in the extent of the treatment to be applied. The physician is often confronted with a mild hyperlipidemia in a young adult (cholesterol and/or triglycerides around 260 mg/dl for instance) and may be hesitant to prescribe a drastic change in his dietary habits for a lifetime. The presence in the family of a brother of 40 years of age with angina pectoris, of a father who died of a myocardial infarction before 55, should incite the physician to take this plasma lipid abnormality seriously.

c) Age and sex of the patient. One should be hesitant to administer drugs to children unless the short term risks of the disease are such that they far outweigh any potential hazard from long term drug administration. Such is the case for the homozygous form of familial hypercholesterolemia where premature CHD and death are the rule.

Until menopause women are definitely less likely to develop CHD than men even if they have severe hypercholesterolemia (44). The physician must be cautious however in postponing treatment in women with FH since premature atherosclerosis may develop in some of them before the menopause and it is difficult to make any prediction in an individual case although the familial context again may help in this regard.

d) Socio-economic setting. On occasion the resistance to treatment with diet or drug will come not from the disease itself, but either from the own will of the patients or from his own socio-cultural constraints. Obviously the patient's collaboration and his ability to comply are essential for any treatment to be successful. Psychological reasons (food dependence, fear of drugs, alcoholism ...) personal beliefs (food faddism, religious taboos ...) or simple economic imperatives might preclude any attempts at modifying dietary habits or administering a drug. Motivation is needed to follow a treatment and it might happen that the same patient, who told his physician he would prefer a natural sudden death to a joyless life away from his gourmet club, would be back to seek treatment after his first myocardial infarction.

When the time comes to prescribe a lipid-lowering drug it is most important to let the patient know that it is to be added to the diet and not to be a substitute for the diet. This may sound naive and obvious to the physician but it is not for the patient who would rather assume spontaneously that the drug will dispense him from following a diet. If the physician takes care to weigh the patient under standard conditions at each visit, as an essential part of the follow-up, he may pick up an increase in weight after initiation of drug treatment, giving him a clue that the diet is being neglected.

One must always start with the most potent drug having the least side effects. Proceed by trial and error adding or substituting other drugs according to response until the most efficient combination is found. Undesirable side effects are to be monitored very closely especially in the first few months of treatment. The results must be ascertained not only from the changes in plasma lipids but also from changes in size of the xanthomas remembering that in an occasional subject xanthomas may regress before any lowering in plasma lipids is achieved (45). When present the symptomatology of atherosclerotic vascular disease should be closely followed and any evidence of improvement or worsening recorded. In all instances the benefits of drug treatment should be carefully weighed against potential hazards.

SPECIFIC DRUGS

When the step of drug administration is reached (6,7,12,41,46), treatment should be initiated with one of the first-line drugs (Table 5). Cholestyramine (or one of its analogues such as colestipol), an insoluble ion exchange resin, and VLDL-cholesterol but will often tend to increase LDL-cholesterol although most of the time not reaching abnormal levels. We have been able to demonstrate its effectiveness at reducing LDL-cholesterol up tp 30% in about 1/3 of cases of familial hypercholesterolemia, usually in females with a high triglyceride/cholesterol ratio (type IIa) (48,49). In contrast to these responders where the HDL-cholesterol is raised by clofibrate (50), about 1/3 of FH patients are resistant and may even show an increase in LDL-cholesterol while HDL-cholesterol is unaffected. They are mostly men with a IIb phenotype. Thus clofibrate should be discontinued whenever such an effect is noticed whether in FH or in FCH. The failure of the Coronary Drug Project to demonstrate any improvement in CHD mortality (51) might have been related to the large proportion of resistant cases in the sampling (52) since the overall effect of clofibrate on plasma cholesterol was rather small (6.5% fall). Clofibrate may potentiate the effect of coumarin anticoagulants (46) and its lipid-lowering effect may be opposed by some oral contraceptive drugs (53). Its major side effects (51) are cholelithiasis, loss of libido and, in rare instances, a myalgic syndrome (54) which is accompanied by an elevation of plasma creatinine phosphokinase (CPK). If this reversible syndrome characterized by tiredness, pain, stiffness and cramps in the muscles of the limbs develops, the drug should be stopped. If the symptoms are milder, halving the dose might be sufficient to alleviate them. The new and more potent derivatives of clofibrate currently investigated are promising. One of them ciprofibrate, that we have been studying recently, appears to be a definite improvement over clofibrate as far as potency, LDL-lowering effect and dosage schedule are concerned. Clofibrate and most of its analogues cause an increase in liver peroxysomes (55) in rats and although this appears to be a species-specific effect it has raised some concerns about a potential carcinogenicity of these drugs. In human studies the WHO trial (56) has reported an increased incidence of cancer in the clofibrate treated group although it was not apparent in all three centers where the study was conducted. The CDP trial (51) on the other hand failed to show any difference between the 3 major study groups in cancer mortality.

Probucol is another well tolerated drug which shares the indications of cholestyramine (FH and FCH). It is administered twice a day as clofibrate but should be given with the meal for a greater absorption and efficacy. Its sulfur-containing highly lipolytic structure is unique among lipid lowering agents. It has few side effects, mainly loose stools to which some patients do not object because it improves their constipation. Its effect is usually sustained with rare escape phenomena. In a double blind crossover trial carried out at our lipid clinic in FH patients, probucol lowered plasma cholesterol an average of 13.5% (57). This effect was mostly ascribed to a fall in LDL-cholesterol. It

Table 5. FIRST-LINE LIPID-LOWERING DRUGS

Drug and dosage	Effect on plasma lipoproteins			Indications (hyperlipidemia)
	VLDL (TG)	IDL (TG, CH)	LDL (CH)	
CHOLESTYRAMINE, 4g (16-24 g/d)	0	0	+++	FH,FCH, (IIa or IIb)
CLOFIBRATE, 0.5g (2 g/d)	++	+++	++ *	FDL,FHPB (FCH, FH IIa, FH IIb, FMH)
PROBUCOL, 0.25g (1 g/d)	0	0	++	FH,FCH
NICOTINIC ACID 0.1,0.5g (1-4 g/d)	+++	+++	++	FMH,FDL,FHPB,FH

Abbreviations: VLDL= very low density lipoproteins, IDL= intermediate density lipoproteins; LDL= low density lipoproteins; TG= triglycerides; CH= cholesterol; FH= familial hypercholesterolemia; FCH= familial combined hyperlipidemia; FDL= familial dysbetalipoproteinemia; FHPB= familial hyperprebetalipoproteinemia; FMH= familial mixed hyperlipidemia; Roman numerals= lipoprotein phenotype; 0= absence of effect; + = small effect; ++ = moderate effect; +++ = strong effect (see text)
* In the responders only.

promotes the excretion of cholesterol from the body by preventing the reabsorption of bile acids (47). It is effective in reducing the atherogenic low density lipoproteins and is thus indicated in familial hypercholesterolemia as well as in familial combined hyperlipidemia where a reduction of plasma cholesterol is needed. It tends to induce constipation and many patients object to its consistency (gritty feeling in the mouth and throat) and bulk but it may cause up to a 40% reduction in plasma cholesterol. It may also induce a more or less transient increase in plasma triglycerides.

Clofibrate (or any of its derivatives such as procetofene or bezafibrate) has a major effect on triglycerides and VLDL-cholesterol. It is very effective at reducing intermediate density lipoproteins (IDL) so that it is the drug of choice in familial dysbetalipoproteinemia (type III). In familial hypertriglyceridemia (type IV), it will reduce both triglycerides and VLDL-cholesterol but will often tend to increase LDL-cholesterol although most of the time not reaching abnormal levels. We have been able to demonstrate its effectiveness at reducing LDL-cholesterol up tp 30% in about 1/3 of cases of familial hypercholesterolemia, usually in females with a high triglyceride/cholesterol ratio (type IIa) (48,49). In contrast to these responders where the HDL-cholesterol is raised by clofibrate (50), about 1/3 of FH patients are resistant and may even show an increase in LDL-cholesterol while HDL-cholesterol is unaffected. They are mostly men with a IIb phenotype. Thus clofibrate should be discontinued whenever such an effect is noticed whether in FH or in FCH. The failure of the Coronary Drug Project to demonstrate any improvement in CHD mortality (51) might have been related to the large proportion of resistant cases in the sampling (52) since the overall effect of clofibrate on plasma cholesterol was rather small (6.5% fall). Clofibrate may potentiate the effect of coumarin anticoagulants (46) and its lipid-lowering effect may be opposed by some oral contraceptive drugs (53). Its major side effects (51) are cholelithiasis, loss of libido and, in rare instances, a myalgic syndrome (54) which is accompanied by an elevation of plasma creatinine phosphokinase (CPK). If this reversible syndrome characterized by tiredness, pain, stiffness and cramps in the muscles of the limbs develops, the drug should be stopped. If the symptoms are milder, halving the dose might be sufficient to alleviate them. The new and more potent derivatives of clofibrate currently investigated are promising. One of them ciprofibrate, that we have been studying recently, appears to be a definite improvement over clofibrate as far as potency, LDL-lowering effect and dosage schedule are concerned. Clofibrate and most of its analogues cause an increase in liver peroxysomes (55) in rats and although this appears to be a species-specific effect it has raised some concerns about a potential carcinogenicity of these drugs. In human studies the WHO trial (56) has reported an increased incidence of cancer in the clofibrate treated group although it was not apparent in all three centers where the study was conducted. The CDP trial (51) on the other hand failed to show any difference between the 3 major study groups in cancer mortality.

Probucol is another well tolerated drug which shares the indications of cholestyramine (FH and FCH). It is administered twice a day as clofibrate but should be given with the meal for a greater absorption and efficacy. Its sulfur-containing highly lipolytic structure is unique among lipid lowering agents. It has few side effects, mainly loose stools to which some patients do not object because it improves their constipation. Its effect is usually sustained with rare escape phenomena. In a double blind crossover trial carried out at our lipid clinic in FH patients, probucol lowered plasma cholesterol an average of 13.5% (57). This effect was mostly ascribed to a fall in LDL-cholesterol. On the other hand it did not change plasma triglycerides significantly. The effect of probucol was additive to that of the diet so that the average fall in plasma cholesterol for the diet responders was 25%. Among its drawbacks it has a long half-life, tends to accumulate in adipose tissue and it lowers HDL-cholesterol. Since both LDL-C and HDL-C are lowered in responders, the LDL/HDL ratio is not adversely affected. In addition the fact that xanthelasmas regress in some patients treated with probucol tends to indicate that the lowered HDL concentrations do not prevent cholesterol from coming out of the tissues, a rather reassuring observation. Finally probucol has been effective in FH patients who had been found resistant to the cholesterol-lowering effect of clofibrate (58). Its exact mechanism of action is not known with certainty, but it appears to influence cholesterol catabolism by promoting bile acid excretion (59) and increase LDL-apolipoprotein B fractional removal (60), it may also have an effect on cholesterol synthesis (61).

Although nicotinic acid and its derivative may produce major side effects (51), it is included among first line drugs because of the large experience accumulated over the years (50,62), its relative safety, and its effectiveness where other drugs have failed. As clofibrate it may lower all three major atherogenic lipoproteins (LDL, IDL, VLDL) and is especially effective in familial mixed hypertriglyceridemia (type V). It is a good drug to give in familial hypertriglyceridemia (type IV) where clofibrate has failed and it has been used with success in FH even in the homozygous patient. The dose should be built up gradually until a daily administration of 2 to 4 g is reached. Side effects include flushing and redness of the face and trunk presumably related to prostaglandin release and alleviated by acetylsalicylic acid. Nausea, diarrhea, abdominal pain, dizziness, pollakyuria and dysuria have been reported. Abnormalities of liver function tests, hyperuricemia and hyperglycemia may develop. Its administration may aggravate diabetes or peptic ulcers, provoque an attack of gout or gastritis. Thus side effect monitoring is especially needed. It has many different sites of action to account for its lipid-lowering effect such as inhibition of lipolysis, reduction of triglyceride, VLDL and cholesterol synthesis and enhancement of chylomicrons, VLDL and cholesterol catabolism.

Some combinations of first line drugs appear to be especially successful (12). Cholestyramine in particular may be advantageously combined with clofibrate (adds a triglyceride lowering effect), probucol (helps control constipation, enhances the LDL-lowering effect) or nicotinic acid (very effective combination on all lipoprotein fractions). Under special circumstances, it may be needed to resort to second line drugs (Table 6) either singly or in combination. These are set

Table 6. SECOND-LINE LIPID-LOWERING DRUGS

Drug (dosage)	Effect on plasma lipoproteins			Indications (hyperlipidemia)
	VLDL (TG)	IDL (TG, CH)	LDL (CH)	
DEXTRO-THYROXINE 2 & 4 mg (2-8 mg/d)	+	++	++	FH,FDL,FHPB
^-SITOSTEROL 3g (3-18 g/d)	0	++	++ *	FH
NEOMYCIN 0.5g (1.5-2.0 g/d)	0	?	++	FH
PARAAMINO-SALICYLIC ACID 0.5g (6-8 g/d)	+	?	++	FH (IIa, IIb)
NORETHINDRONE 5mg (5 mg/d)	++	?	0	FMH
OXANDROLONE 2.5mg (7.5 mg/d)	++	++	0	FDL,FHPB,FMH

Abbreviations as per table 5.

apart because of either their side effects, their narrow indications or their low efficacy. They should be prescribed for specific indications, used in combination with the first line drugs or be given when none of the former may be administered (6,7,12). Dextrothyroxin (63) has a strong LDL-lowering effect, but may worsen angina pectoris or induce arrhythmias if not given with betablockers (64). Betasitosterol (65) is a well tolerated mild lipid-lowering agent potentially lithogenic. Neomycin has the indications of cholestyramine but may affect adversely the intestinal flora and cause nausea and diarrhea (66). Paraaminosalicylic acid (PAS) has many gastro-intestinal side effects (67). Norethindrone acetate (68) and oxandrolone (69) stimulate the activity of lipoprotein lipase and may be used in familial mixed hypertriglyceridemia (type V) respectively in females and in males.

Fifth step: Extreme measures

When the dietary and pharmacological managements fail to achieve satisfactory lipid-lowering in an obviously progressive and life threatening disease, one may have recourse to more extreme measures. These are usually practiced in specialized centers, they may have major drawbacks and several are still at an experimental stage. Partial ileal bypass surgery (70) has been used in various forms of hyperlipidemia but we tend to restrict its use to resistant cases of heterozygous FH patients. Porta-caval shunting (71) is still considered a rather unpredictable experimental procedure for severe cases of homozygous FH. Its effectiveness could be enhanced by performing in addition a partial ileal bypass (72). Ligature of the chyliferous vessels near the renal pedicle (73) is another surgical procedure which is not extensively used. Repeated plasmaphereses have been applied with some success to the treatment of FH (74) and could eventually be replaced by the simpler technique of extracorporeal retrieval of LDL on heparin-agarose beads (75).

Summary and conclusions

A systematic approach to the treatment of primary hyperlipidemias must begin with an accurate diagnosis. At this step the physician should give careful attention to the multitude of factors likely to affect plasma lipid levels and look for clinical clues which will be useful in establishing the presence and the nature of the plasma lipid disorder. He should refer to a practical classification of hyperlipidemias and keep in mind that more and more attention should be given to the apoprotein moiety of the plasma lipoproteins. When the nature and severity of the disease are ascertained, the decision to treat must rest on a careful assessment of the clinical and familial context. Since in most instances the ultimate goal is to prevent complications from atherosclerosis, the other risk factors for coronary heart disease should be dealt with simultaneously. The dietary approach precedes the use of drugs. It should be systematic, based on an extensive survey of the dietary habits and take into account the current knowledge which relates dietary factors to the pathogenesis of atherosclerosis. Drugs should be used when dietary management has not provided a "satisfactory response" but at all time it should be individualized. Often one is compelled to proceed by trial and error until the most effective combination of drugs is found, making a judicious choice among the first and second line drugs. One may resort to more extreme measures, surgical or other, when all else fails in severely resistant cases. At all times treatment must be adapted to the type of plasma lipoprotein abnormality and the benefits of treatment cautiously weighed against potential hazards. The physician must feel secure that his efforts will be rewarded if he perseveres. There now is enough evidence in the medical literature to think that a large and sustained reduction in plasma lipids may, over a number of years, not only arrest the atherogenic process but promote its regression.

REFERENCES

1. J. Davignon, Current views on the etiology and pathogenesis of atherosclerosis. In: "Hypertension: Physiopathology and Treatment", J. Genest, E. Koiw and O. Küchel, eds., McGraw-Hill Book Co. , New York (1977).
2. M. L. Armstrong, Evidence of regression of atherosclerosis in primates and man. Postgrad. Med. J. 52:456 (1976).
3. C. R. Ost, and S. Stenson, Regression of peripheral atherosclerosis during therapy with high doses of nicotinic acid. Scand. J. Clin. Lab. Invest. Suppl. 99:241 (1967).
4. L. L. Basta, C. Williams, J. M. Kioschos, and A. Spector, Regression of atherosclerotic stenosing lesions of the renal arteries and spontaneous cure of systemic hypertension through control of hyperlipidemia. Amer. J. Med. 86:139 (1976).
5. R. Barndt Jr, D. H. Blankenhorn, D. W. Crawford, and S. H. Brooks, Regression and progression of early femoral atherosclerosis in treated hyperlipoproteinemic patients. Ann. Int. Med. 86:139 (1977).
6. J. Davignon, The hyperlipoproteinemias. In: "Current Therapy", H.F. Conn, ed., W. B. Saunders, Philadelphia (1978).
7. J. Davignon, and S. Lussier-Cacan, Approche systématique au traitement des hyperlipidémies héréditaires. Rev. Med. Liège 35:286 (1980).
8. A. Gattereau, E. Elie, W. Figareau, and J. Davignon, Etude preliminaire des caracteristiques du lipidogramme liees a la presence de l'anneau corneen chez un groupe d'Haitiens. Union Med. Can. 104:1359 (1975).
9. R. H. Rosenman, R. J. Brand, R. I. Sholtz, and C.D. Jenkins, Relation of corneal arcus to cardiovascular risk factors and the incidence of coronary heart disease. New Engl. J. Med. 291:1322 (1974).
10. Ph. Douste-Blazy, Y.L. Marcel, L. Cohen, J.M. Giroux, and J. Davignon, Increased frequency of apo E-ND phenotype and hyperapobetalipoproteinemia in normolipidemic subjects with xanthelasmas of the eyelids. Ann. Int. Med. 96:164 (1982).

11. A. Watanabe, A. Yoshimura, T. Wakagushi, R. Tatami, K. Veda, R. Veda, T. Haba, T. Kametani, J. Koizumi, S. Ito, W. Ohta, S. Miyamoto, H. Mabuchi, and R. Takeda, Serum lipids, lipoprotein lipids and coronary heart disease in patients with xanthelasma palpebrarum. Atherosclerosis 38:283 (1981).
12. J. Davignon, Les médicaments dans le traitement des hyperlipidémies - nécessités et limites. In: "Symposium sur le Probucol", J. Beaumont and B. Jacotot, eds., Excerpta Medica, Amsterdam (1981).
13. J. Davignon, S. Lussier-Cacan, A. Gattereau, P. P. Moll, and C. F. Sing, Studies of a large French-Canadian kindred with two forms of hyperlipoproteinemia. In: "Atherosclerosis V", A. M. Gotto Jr., L. C. Smith and B. Allen, eds., Springer-Verlag, New York (1980).
14. D. S. Fredrickson, J. L. Goldstein, and M. S. Brown, The familial hyperlipoproteinemia. In: "The metabolic basis of inherited disease", J. B. Stanbury, J. B. Wyngaarden and D. S. Fredrickson, eds., McGraw-Hill, New York (1978).
15. P. Elias, and L. A. Goldsmith, Intertriginous xanthomata in type 2 hyperbetalipoproteinemia. Arch. Dermatol. 107:761 (1973).
16. D.S. Fredrickson, R.I. Levy, and R. S. Lees, Fat transport in lipoproteins - an integrated approach to mechanisms and disorders. New Engl. J. Med. 276:94,148,215,273, (1967).
17. J. L. Goldstein, W.R. Hazzard, H.G. Schrott, E.L. Bierman, and A.G. Motulsky, Hyperlipidemia in coronary heart disease. II. Genetic analysis of lipid levels in 176 families and delineation of a new inherited disorder, combined hyperlipidemia. J. Clin. Invest. 52:1544 (1973).
18. J. D. Brunzell, H.G. Schrott, A.G. Motulsky, and E.L. Bierman, Myocardial infarction in the familial forms of hypertriglyceridemia. Metabolism 25:313 (1976).
19. C. J. Glueck, R.W. Fallat, R. Millet, and P. M. Steiner, Familial hyperalphalipoproteinemia: studies in eighteen kindreds. Metabolism 24: 1243 (1975).
20. F. T. Hatch, and R. S. Lees, Practical methods for plasma lipoprotein analysis. In: "Advances in Lipid Research", R. Paoletti and D. Kritchevsky, eds., Academic Press, New York (1968).
21. J. Davignon, and M. Langelier, Electrophorèse et ultracentrifugation combinées dans le diagnostic des hyperlipidémies primaires. Union Méd. Can. 100:2120 (1971).
22. D. S. Fredrickson, J. Morganroth, and R. I. Levy, Type III hyperlipoproteinemia: an analysis of two contemporary definitions. Ann. Int. Med. 82:150 (1975).
23. B. H. Masket, R.I. Levy, and D. S. Fredrickson, The use of polyacrylamide gel electrophoresis in differentiating type III hyperlipoproteinemia. J. Lab. Clin. Med. 81:794 (1973).

24. M. D. Curry, W.J. McConathy, P. Alaupovic, J. H. Ledford, and M. Popovic, Determination of human lipoprotein E by electroimmunoassay. Biochim. Biophys. Acta 439:413 (1976).
25. G. Utermann, M. Hees, and A. Steinmetz, Polymorphism of apolipoprotein E and occurrence of dysbetalipoproteinemia in man. Nature 269:604 (1977).
26. J. L. Breslow, and V. I. Zannis, Biochemical and genetic studies of human apolipoprotein E. In: "Drugs affecting lipid metabolism", R. Fumagalli, K. Kritchevsky and R. Paoletti, eds., Elsevier North-Holland, Amsterdam (1980).
27. J. C. La Rosa, R. I. Levy, P. Herbert, S. E. Lux, and D. S. Fredrickson, A specific apoprotein cofactor for lipoprotein lipase. Biochem. Biophys Res. Communic. 41:57 (1970).
28. W. C. Breckenridge, J. A. Little, G. Steiner, A. Chow, A. and M. Poapst, Hypertriglyceridemia associated with deficiency of apolipoprotein CII. New Engl. J. Med. 298:1265 (1978).
29. W. C. Breckenridge, D. Cox, and J. A. Little, Apolipiprotein CII deficiency. In: Atherosclerosis V. A. M. Gotto Jr., L. C. Smith and B. Allen eds., Springer-Verlag, New York (1980).
30. J. A. Little, D. Cox, D., W. C. Breckenridge, and V. M. McGuire, Introduction to deficiencies in apolipoproteins CII and EIII with some associated clinical findings. In: Atherosclerosis V. A. M. Gotto Jr., L. C. Smith and B. Allen eds., Springer-Verlag, New York (1980).
31. P. O. Kwiterovich, Jr., P. S. Bachorik, H. H. Smith, V. A. McKusick, W. E. Connor, B. Teng, and A. D. Sniderman, Hyperapobetalipoproteinaemia in two families with xanthomas and phytosterolaemia. Lancet 1:466 (1981).
32. A. D. Sniderman, S. Shapiro, D. Marpole, B. Skinner, B., B. Teng, and P. O. Kwiterovich, Jr., Association of coronary atherosclerosis with hyperapobetalipoproteinemia (increased protein but normal cholesterol levels in human plasma low density (β) lipoproteins). Proc. Natl. Acad. Sci. USA, 77:604 (1980).
33. A. K. Bhattacharyya, and W. E. Connor, β-sitosterolemia and xanthomatosis, a newly described lipid storage disease in two sisters. J. Clin. Invest. 53:1033 (1974).
34. C. J. Glueck, P. Gartside, P., R. W. Fallat, J. Sielski, and P. M. Steiner, Longevity syndromes: Familial hypobeta and familial hyperalphalipoproteinemia. J. Lab. Clin. Med. 88:941 (1976).
35. J. L. De Gennes, G. Turpin, G. and J. Truffert, Dépistage et identification des hyperlipidêmies idiopathique. Un nouveau test. Comparaison des concentrations des lipides circulants à la 8e et à la 12e heure de jeûne. Nouv. Presse Méd. 1:1629 (1972).

36. J. Morganroth, R. I. Levy, A. E. McMahon, and A. J. Gotto, Jr, Pseudohomozygous type II hyperlipoproteinemia. J. Pediat. 85:639 (1974).
37. A. K. Bhattacharyya, and W. E. Connor, Familial diseases with storage of sterols other than cholesterol (cerebro-tendinous xanthomatosis and β-sitosterolemia and xanthomatosis). In: "The metabolic basis of inherited disease", J.B. Stanbury, J.B. Wyngaarden and D.S. Fredrickson, eds., McGraw-Hill, New York (1978).
38. R. H. Grimm Jr, A. S. Leon, D. B. Hunninghake, K. Lenz, P. Hannan, and H. Blackburn, Effects of thiazine diuretics of plasma lipids and lipoproteins in mildly hypertensive patients. Ann. Int. Med. 94:7 (1981).
39. P. Bielmann, G. Leduc, and J. Davignon, Bêta-bloquants sélectif (métoprolol) vs non-sélectif (propranolol): effets comparés sur la composition des lipoprotéines plasmatiques et la pression sanguine. Union Méd. Can. 109:1734 (1980).
40. W. E. Connor, and W. L. Connor, Dietary treatment of hyperlipo-proteinemia. In: "Hyperlipidemia, diagnosis and therapy", Grune and Stratton, New York (1977).
41. J. Davignon, S. Lussier-Cacan, and J. Ph. Wolf, Le traitement de l'athérogenèse, les facteurs de risque, en particulier les hyperlipidémies. In: "L'athérogenèse", J. Hugues, ed., Masson, Paris (1979).
42. J. Davignon, N. Leboeuf, and S. Lussier-Cacan, Aspects nutritionnels du traitement des hyperlipidémies et de l'athérosclérose. Union Méd. Can. 109:656 (1980).
43. A. J. Vergroesen, Physiological effects of dietary linoleic acid. Nutr. Rev. 35:1 (1977).
44. J. Slack, Inheritance of familial hypercholesterolemia. Ather. Rev. 5:35 (1979).
45. A. J. Palmer, and R. Blacket, Regression of xanthomata of the eyelids with modified fat diet. Lancet I:66 (1972).
46. D. B. Hunninghake, and J. L. Probstfield, Drug treatment of hyperlipoproteinemia. In: "Hyperlipidemia, diagnosis and therapy", B.M. Rifkind and R.I. Levy, eds., Grune et Stratton, New York (1977).
47. S. M. Grundy, Treatment of hypercholesterolemia by interference with bile acid metabolism. Arch. Int. Med. 130:638 (1972).
48. J. Davignon, F. Aubry, C. Noël, Y. Lapierre, and M. Lafortune, Heterogeneity of familial hyperlipoproteinemia type II on the basis of fasting plasma triglyceride/cholesterol ratio and plasma cholesterol response to chlorophenoyisobutyrate. Rev. Can. Biol.30:307 (1971).

49. J. Davignon, A. Sniderman, M. N. Cayen, W. T. Robinson, M. Kraml, E. Reichel, and S. Lussier-Cacan, Sensitivity and resistance to the cholesterol-lowering effect of clofibrate - Pharmacokinetic studies, plasma apolipoprotein B levels and therapeutic implications. In: "Drugs affecting lipid metabolism", R. Fumagalli, D. Kritchevsky and R. Paoletti, eds., Elsevier/North Holland, Amsterdam (1980).
50. D. B. Hunninghake, Drug treatment of type II hyperlipoproteinemia - Effects on plasma lipid and lipoprotein levels. In: "Atherosclerosis V", A. M. Gotto Jr., L. C. Smith and B. Allen, eds., Springer-Verlag, New York (1980).
51. The Coronary Drug Project Research Group: Clofibrate and niacin in coronary heart disease. J. Amer. Med. Assoc. 231:360 (1975).
52. R. S. Lees, Clofibrate and atherosclerosis. New Engl. J. Med. 300:491 (1979).
53. R. B. W. Smith, and I. A. M. Prior, Oral contraceptive opposition to hypocholesterolaemic action of clofibrate. Lancet 1:750 (1968).
54. T. Langer, and R. I. Levy, Acute muscular syndrome associated with administration of clofibrate. New Engl. J. Med. 279:856 (1968).
55. J. K. Reddy, D.L. Azarnoff, and C. E. Hignite, Hypolipidaemic hepatic peroxisome proliferators form a novel class of chemical carninogens. Nature 283:397 (1980).
56. M. F. Oliver, J. A. Heady, J. N. Morris, and Cooper, J, A cooperative trial in the primary prevention of ischaemic heat disease using clofibrate - Report from the committee of principal investigators. Brit. Heart J. 40:1069 (1978).
57. J. LeLorier, S. Dubreuil-Quidoz, S. Lussier-Cacan, Y. S. Huang, and J. Davignon, Diet and probucol in lowering cholesterol concentrations. Arch. Int. Med. 137:1429 (1977).
58. M. Enjalbert, S. Lussier-Cacan, S. Dubreuil-Quidoz, J. LeLorier, and J. Davignon, Usefulness of probucol in treating primary hypercholesterolemia. Can. Med. Assoc. J. 123:754 (1980).
59. T. A. Miettinen, Mode of action of a new hypocholesteraemic drug (DH-581) in familial hypercholesteraemia. Atherosclerosis 15:163 (1972).
60. P. J. Nestel, and T. Billington, Effects of probucol on low density lipoprotein removal and high density lipoprotein synthesis. Atherosclerosis 38:203 (1981).
61. Y. S. Huang, M. Asselin, C. Gianoulakis, M. Lis, M. Chrétien, and J. Davignon, Liver lipid metabolism and effect of cholesterol-lowering agents in the hyperlipidemia induced in rats by implantation of the pituitary derived MtT-F_4 tumor. Can. J. Physiol. Pharmacol. 58:767 (1979).

62. L. A. Carlson, and L. Oro, Effect of treatment with nicotinic acid for one month on serum lipids in patients with different types of hyperlipidemia. Atherosclerosis 18:1 (1973).
63. L. A. Simons, and N. B. Myant, The effect of D-thyroxine on the metabolism of cholesterol in familial hyperbetalipoproteinaemia. Atherosclerosis 19:103 (1974).
64. D. M. Krikler, D. Lefevre, and B. Lewis, Dextrothyroxine with propranolol in treatment of hypercholesterolaemia. Lancet 1:934 (1971).
65. A. M. Lees, H. Y. I. Mok, R. S. Lees, M. A. McCluskey, and S. M. Grundy, Plant sterols as cholesterol-lowering agents: clinical trials in patients with hypercholesterolemia and studies of sterol balance. Atherosclerosis 28:325 (1977).
66. P. Samuel, C. M. Holtzman, and J. Goldstein, Long term reduction of serum cholesterol levels of patients with atherosclerosis by small doses of neomycin. Circulation 35:938 (1967).
67. P. J. Barter, W. E. Connor, A. A. Spector, M. Armstrong, L. Sonja, B. S. Connor, and M. A. Newman, Lowering of serum cholesterol and triglycerides by para-aminosalicylic acid in hyperlipoproteinemia. Ann. Int. Med. 81:619 (1974).
68. C. J. Glueck, V. W. Brown, R. I. Levy, H. Greten, and D. S. Fredrickson, Amelioration of hypertriglyceridemia by progestational drugs in familial type V hyperlipoproteinemia. Lancet I:1290 (1969).
69. C. J. Glueck, Effects of oxandrolone on plasma triglycerides and postheparin lipolytic activity in patients with type III, IV and V familial hyperlipoproteinemia. Metabolism 20:691 (1971).
70. H. Buchwald, R. Moore, and R. L. Varco, Surgical treatment of hyperlipidemia. Circulation 49, Suppl. 1:1 (1974).
71. T. E. Starzl, C. W. Putnam, and L. J. Koep, Portacaval shunt and hyperlipidemia. Arch. Surg. 113:71 (1978).
72. T. A. Miettinen, Comparison of cholestyramine, ileal-bypass and portacaval shunt in the treatment of familial hypercholesterolemia. In: "Atherosclerosis V". A. M. Gotto Jr., L. C. Smith and B. Allen, eds., Springer-Verlag, New York (1980).
73. M. Servelle, Ligation of the chyliferous vessels in the treatment of arteriosclerosis. In: "International Conference on Atherosclerosis", L. A. Carlson, R. Paoletti, C. R. Sirtori and G. Weber, eds., Raven Press, New York (1978).
74. G. R. Thompson, R. Lowenthal, and N. B. Myant, Plasma exchange in the management of homozygous familial hypercholesterolaemia. Lancet I:1208 (1975).
75. P. J. Lupien, S. Moorjani, and J. Awad, A new approach to the management of familial hypercholesterolaemia: removal of plasma-cholesterol based on the principle of affinity chromatography. Lancet I:1261 (1976).

LIPO AND APOPROTEIN ASSOCIATIONS IN CORONARY ARTERY DISEASE

H. Van-Zeller

Biochemistry Department of Faculty of Pharmacy

Av. Forças Armadas - 1699 Lisboa Codex Portugal

ABSTRACT:

Triglycerides and cholesterol concentrations in total plasma and lipoprotein fractions, obtained by preparative ultracentrifugation, were determined in 104 middle aged men subjected to coronary angiography for suspected myocardial ischaemia. Plasma apoprotein concentrations were measured by immunoelectrophoresis. The degree, number and length of stenoses in major coronary arteries were scored for each angiogram.

Patients inthe bottom and top quintiles of the angiogram score, were compared. Significant differences were found in LDL, in total HDL and in HDL_2 cholesterol concentration. The latter was considerably higher in the bottom quintile subjects ($p< 0.005$). No differences were found in the cholesterol content of HDL_3 and lipoproteins of density < 1.019, in plasma triglyceride and apoproteins A-I, A-II or E. HDL_2/LDL cholesterol concentration ratio was found to be the best discriminator between the two groups of subjects.

INTRODUCTION

Angiographic studies have demonstrated that the relationship of lipoprotein levels to clinical disease reflects underlying correlations with the severity of coronary atherosclerosis. There is a relative paucity of information, however, on the relationship of coronary atherosclerosis to plasma lipoprotein subclasses and apoproteins.

We tried to improve the assessment of coronary heart disease

Abbreviations: VLDL - very low density lipoproteins
IDL - Intermediate density lipoproteins
LDL - low density lipoproteins
HDL - high density lipoproteins

risk, in individual subjects, and provide insight into the precise nature of the role of plasma lipoprotein metabolism in the atherogenic process. For that purpose, we have quantified the severity of coronary atherosclerosis, in middle aged men, by coronary angiography and have compared the results obtained for various plasma lipoprotein measurements in subjects with severe coronary disease and in those with minimal or absent disease.

Subjects and methods:

The population base consisted of consecutive patients aged 35 to 65 years undergoing coronary angiography for suspected ischaemic heart disease.

The exclusion criteria were females, lipid lowering therapy, previous coronary artery surgery, valvular heart disease, diabetes mellitus and recovery of cholesterol on plasma lipoprotein fractionation of the less than 95%.

The total number of subjects admitted to this study was 104.

The severity of coronary atherosclerosis was quantified by assigning to the coronary angiograms a score reflection the number, degree and severity of stenoses in major segments of the coronary artery tree (right, posterior descending, left main, left anterior descending, circumflex and its obtuse marginal branches). Each stenoses was given a score from 1 to 4 depending upon the degree of narrowing an another for its lenght from 1 to 3. When there was complete obstruction a score of 5 or 6 was given, the former being assigned when there was visible blood flow, distal to the obstruction, via collaterals. The final coronary artery score was then computed as the sum of both scores for all stenoses.

Plasma lipoproteins

All subjects fasted for 14 hours before blood was taken.

A separate aliquot of plasma was centrifuged at a density of 1.019 and the supernatant containing VLDL and IDL was isolated by tube slicing. LDL in the infranatant was then precipitated with heparin (5.000 UI/ml) and manganese chloride (1 M) and the resulting supernatant adjusted to a density of 1.125 with sodium bromide. After further ultracentrifugation the supernatant containing HDL_2 subfraction, was isolated by tube slicing.

Triglyceride in total plasma and in lipoproteins of density < 1.019 were measured enzimatically (Boehringer Manheim).

Cholesterol concentrations in total plasma, density < 1.019 lipoproteins, in HDL and HDL_2, subfraction were measured directly by enzimatic procedures (Boehringer Manheim). Those in LDL and in HDL_3 were measured by difference.

Measurements were also made of the plasma concentrations of apoprotein A-I, A-II, and E. These measurements were made by immunoelectrophoresis using rabbit antisera to purified human apoproteins.

Table 1 : Mean lipoprotein cholesterol ratios of subjects in bottom and top quintile of coronary artery score

	CORONARY ARTERY SCORE		
	Bottom quintile	Top quintile	p
HDL_2 chol./LDL chol.	0.106	0.040	< 0.0025
HDL_2 chol./HDL_3 chol.	0.384	0.170	< 0.05
HDL chol./LDL chol.	0.470	0.350	< 0.05

Table II : Plasma apoprotein concentrations of patients in bottom and top quintile of coronary artery score.

	CORONARY ARTERY SCORE		
	Bottom quintile	Top quintile	P
	(mg/dl)		
Apo A-I	108	110	NS
Apo A-II	38	41	NS
Apo E	5.2	4.9	NS

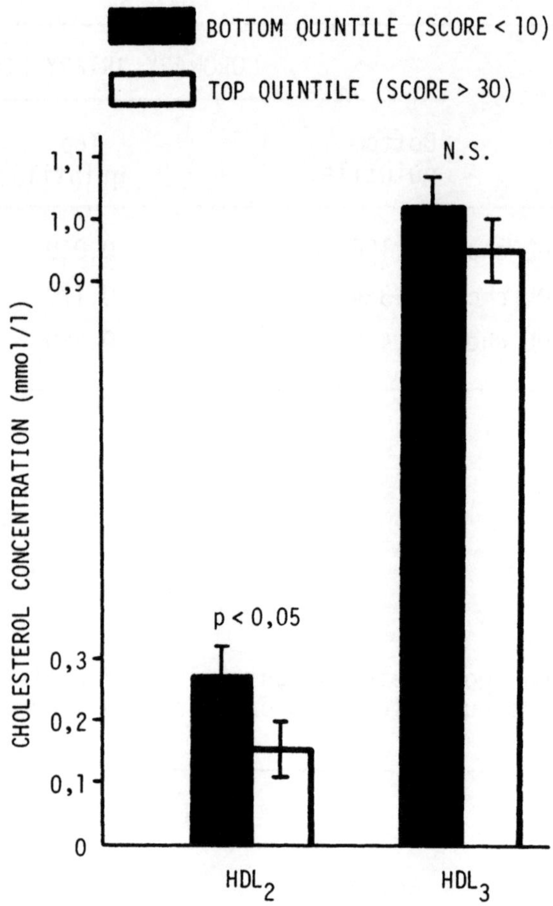

Figure 1. Cholesterol concentrations in total plasma lipoproteins of density < 1.019, LDL and HDL of patients in bottom and top quintile of the coronary artery score.

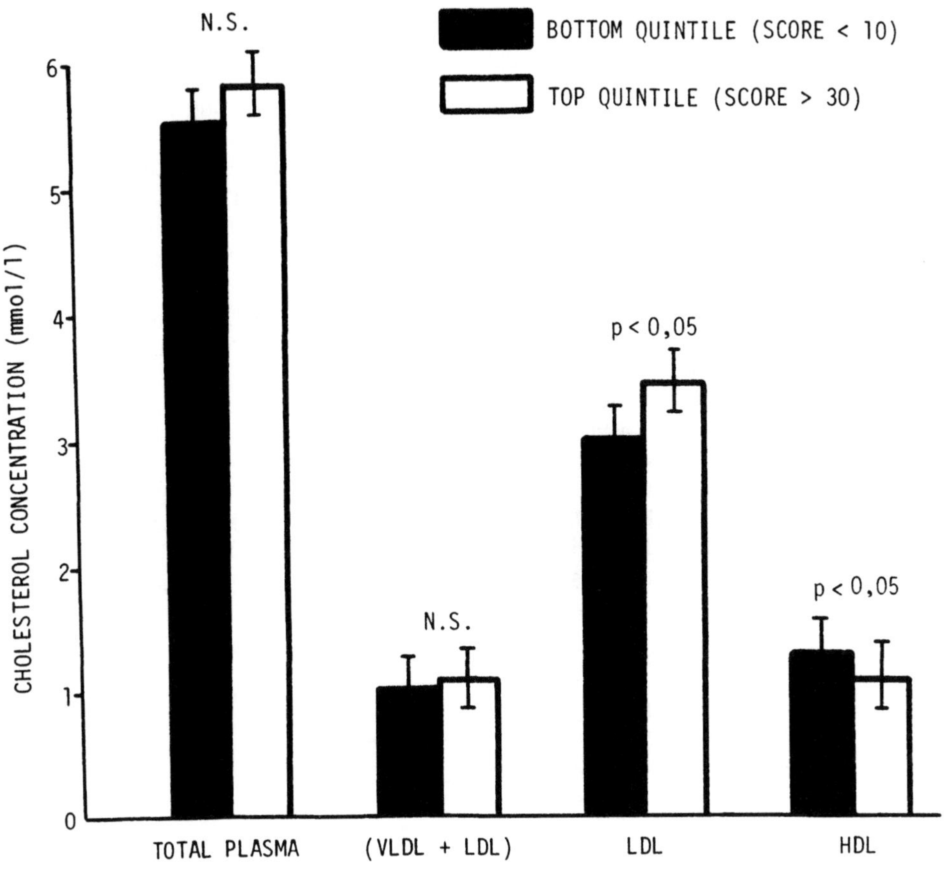

Figure 2. Cholesterol concentrations in HDL_2 and HDL_3 of patients in bottom and top quintile of the coronary artery score.

Results

The data were analysed by comparing subjects in the bottom quintile of coronary artery score with those in the top quintile corresponding to scores of less than 10 and greater than 30. Statiscal analysis were performed by student's test. We verified that there were no significant differences between the two groups of subjects in age, body weight, systolic blood pressure and plasma triglyceride concentration.

The results obtained for cholesterol concentrations are shown in figure 1.

Cholesterol concentration in the whole plasma and lipoproteins density < 1.019 was not significantly different in the two groups of subjects. By contrast, LDL cholesterol concentration was significantly higher and HDL cholesterol concentration significantly lower in subjects in the top quintile of the coronary artery score.

Figure 2 gives the results obtained for HDL subfractions.

The lower concentration of total HDL cholesterol in subjects with severe disease was a reflection of a low HDL_2 cholesterol concentration, this being 44% lower in the severely affected group. On the other hand, the concentration of HDL_3 cholesterol was not significantly different between the two groups.

In an attempt to find a most powerful discriminator between the two groups of patients several ratios were examined and these results are summarized in table 1.

The HDL_2/LDL ratio was the most significantly different between the two groups (62% lower in patients with severe disease). The other ratios were less powerful discriminators.

Table II shows that the apoprotein concentrations were not significantly different in the two groups of subjects.

Discussion

Epidemiological studies have shown that clinical ischaemic heart disease is directly related to LDL cholesterol concentration and inversely related to HDL cholesterol concentration.

The present study demonstrates that the ralationship to HDL cholesterol concentration reflects an underlying relationship to HDL_2 cholesterol concentration.

Further more we found that the most powerful predictor of the severity of coronary atherosclerosis in this group, of middle aged men, was the ratio of HDL_2 to LDL cholesterol concentration.

REFERENCES

1. Castelli WP, Doyle JT, Gordon T, et al. HDL cholesterol and other lipids incoronary heart disease. The Cooperative lipoprotein Phenotyping Study. Circulation 1977;55:767-72.
2. Miller NE,Førde OH, Thelle DS, Mjøs OD. The Tromsø Heart Study. High-density lipoprotein and coronary heart disease: a prospective case-control study. Lancet 1977;i:965-8.

3. Gordon T, Castelli WP, Hjortland MC, Kannel WB, Dawber TR. High density lipoprotein as a protective factor against coronary heart disease. Am. J. Med. 1977;62:707-14.
4. Goldbourt U, Medalie JH. High density lipoprotein cholesterol and incidence of coronary heart disease-the Israeli Ischaemic Heart Disease Study. Am. J. Epidemiol 1979;109:296-308.
5. Hulley SB, Rosenman RH, Bawold RD, Brand RJ. Epidemiology as a guide to clinical decisions: the association between triglyceride and coronary heart disease. N. Engl. J. Med. 1980;302:1383-9.
6. Pearson TA, Bulkley BH, Achuff SC, Kwiterovich PO, Gordis L. The association of low levels of HDL cholesterol and arteriographically defined coronary artery disease. Am J. Epidemiol 1979;109:285-95.
7. Jenkins PJ, Harper RW, Nestel PJ. Severity of coronary atherosclerosis related to lipoprotein concentration. Br Med. J. 1978;ii:388-91.
8. Tan MH, Macintosh W, Weldon KL, Kapoor A, Chandler BM, Hindmarsh, TJ. Serum high density lipoprotein cholesterol in patients with abnormal coronary arteries. Atherosclerosis 1980;37:187-8.
9. Nichols AV. Human serum lipoproteins and their interrelationships. Adv. Biol Med. Phys. 1967;11:109-58.
10. Janus ED, Nicoll A, Wooton R, Turner PR, Magill PJ, Lewis B. Quantitative studies of very low density lipoprotein conversion to low density lipoprotein in normal controls and primary hyperlipidaemic states and the role of direct secretion of low density lipoprotein in heterozygous familial hypercholesterolaemia. Eur. J. Clin Invest. 1980;10:149-59.
11. Patsch JR, Gotto AM Jr, Olivecrona T, Eisenberg S. Formation of high density lipoprotein$_2$-like particles during lipolysis of very low density lipoproteins in vitro. Proc. Natl. Acad. Sci USA 1978;75:4519-23.
12. Havel RJ, Eder HA, Bragdon JH. The distribution and chemical composition of ultracentrifugally separated lipoproteins in human serum. J. Clin. Invest. 1955;34: 1345-53.
13. Albers JJ, Warnick GR, Wiebe D, et al. Multi-laboratory comparison of three heparin-Mn^{2+} precipitation procedures for estimating cholesterol in high-density lipoprotein. Clin. Chem. 1978;24:853-6
14. Alaupovic P. Curry MD, McConarthy WJ. Quantitative determination of human plasma apolipoproteins by electroimmunoassays. In: Carlson LA, Paoletti R, Sirtory CR, Weber G, eds, International conference on atherosclerosis, New York: Raven Press, 1978:109-15.
15. Anonymous. Antihypertensive drugs, plasma lipids and coronary disease. Lancet 1980;ii:19-20.
16. Gotto AM Jr, Miller NE, Oliver MF, eds. High density lipoproteins and atherosclerosis. Amsterdam:Elsevier/North-Holland, 1978.

17. Solberg LA, Enger SC, Hjermann I, et al. Risk factors for coronary and cerebral atherosclerosis in the Oslo study. In: Gotto AM, Smith LC, Allen B, eds. Atherosclerosis V. Berlin: Springer-Verlag, 1980:57-62.
18. Gofman JW, Young W, Tandy R. Ischemic heart disease, atherosclerosis and longevity. Circulation 1966;34:679--97.
19. Albers JJ, Cheung MC, Hazzard WR, High-density lipoproteins in myocardial infarction survivors, Metabolism 1978;27;479-85
20. Gofman JW,DeLalla O, Glazier F, et al. The serum lipoprotein transport system in health, metabolic disorders atherosclerosis and coronary heart disease. Plasma 1954; 2:413-84.
21. Fredrickson DS, Levy RI, Lees RS. Fat transport in lipoproteins-an integrated approach to mechanisms and disorders. N Engl. J. Med. 1967;276:34-44.
22. Zilversmit DB. Atherogenesis: a postprandial phenomenon. Circulation 1979;60:473-85.
23. Ishikawa T, Fidge N, Thelle DS, Førde OH, Miller NE. The Tromsø Heart Study: serum apolipoprotein AI concentration in relation to future coronary heart disease. Eur. J. Clin. Invest. 1978;8:179-82.
24. Avogaro P, Bittolo Bon G, Cazzolato G, Quinci GB. Are apolipoproteins better discriminators than lipids for atherosclerosis? Lancet 1979;i:901-3.
25. Nikkila EA. Metabolic and endocrine control of plasma high density lipoprotein concentration. Relation to catabolism of triglyceride-rich lipoproteins. In: Gotto AM Jr, Miller NE, Oliver MF, eds. High density lipoproteins and atherosclerosis. Amsterdam: Elsevier/ North-Holland, 1978:177-92.
26. Weisgraber KH, Mahley RW. Subfractionation of human high density lipoproteins by heparin-Sepharose affinity chromatography. J. Lipid Res. 1980;21:316-25.
27. Suenram A, McConathy WJ, Alaupovic P. Evidence for the lipoprotein heterogeneity of human plasma high density lipoproteins isolated by three different procedures. In: Lindgren FT, Nichols AV, Krauss RM, eds. High density lipoproteins. Champaign, Illinois: American Oil Chemists Society, 1979:36-41.

CORRELATION OF THE LP-X TEST WITH CLINICAL DIAGNOSIS OF LIVER DISEASES

M.M. Vicente and J.C. Magalhães

Center for Lipid Research
Dept. of Biochem., Fac. Medical Sciences
Campo de Santana, 130 - 1100 Lisbon-PORTUGAL

An abnormal lipoprotein present in blood serum of patients suffering liver diseases was first observed by Russ and al (1).

This lipoprotein was defined by various authors (2, 3) and designated as lipoprotein-X (Lp-x) (4, 5).

The various technics which characterize this lipoprotein have already been described (6, 7, 8).

The first technique used for Lp-X was a precipitation method described by Jirgl (6).

The cathodical migration of Lp-X on agarose gel afforded the development of an imunoelectrophoretic technique*. The appearance of a precipitation line on the cathodical side of application is an indication of Lp-X (fig. 1).

Tallafant and Tovarek (9) described more recently a turbidimetric technique. For us there are a cutting off line at 0,09 D.O. units.

In this paper we compare these three techniques.

The reproductivity of turbidimetric method was tested by dupli cate analysis of 60 sera (table I) and we obtained a good correlation.

In 30 sera we compared the three methods (table II).

We can conclude that the Jirgl test is the less sensitive. The turbidimetric method seems to be a good screening method, using as a cutting off line 0.09 units.

* This technique was comercialized by imuno-diagnostic (Vienna).

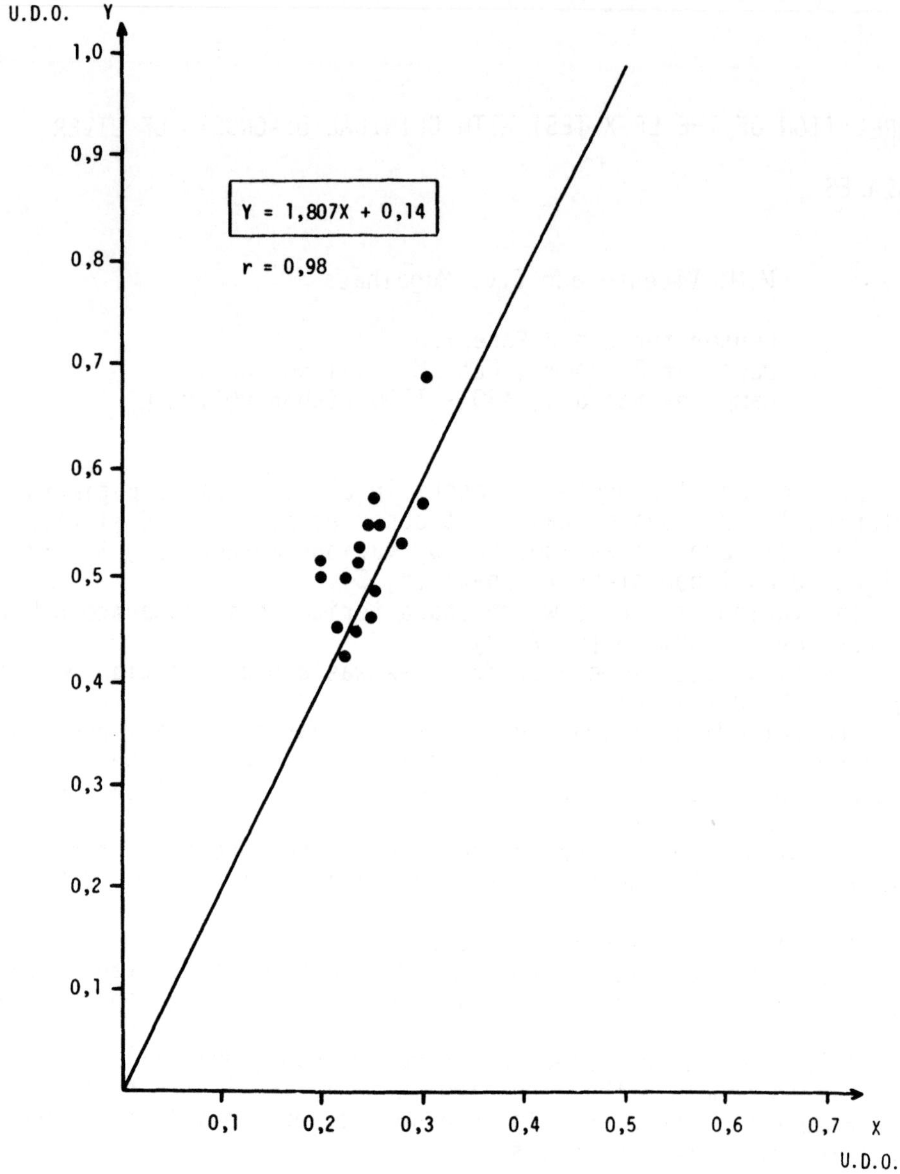

Figure 1: LP-X (++) levels determined by turbidimetry compared with the results of LP-X (+).

TABLE I: Reproductibility of the turbidimetric method

	PATIENTS	r
LP-X(-) Neg.	20	0,951
LP-X(+) Pos.	20	0,980
LP-X(+++) Pos.	20	0,970

TABLE II: Comparison of the three methods

DIAGNOSIS	PATIENTS		JIRGL	TURBIDIMETRY	ELECTROPHORESIS
Cholestasis	16	16	(+)	>0,09 U	LP-X (+)
Hepatic Cirrhosis	12	6	(+)	<0,09 U	LP-X (-)
		6	(-)	<0,09 U	LP-X (-)
Hepatitis s/ Cholestasis	2	2	(-)	<0,09 U	LP-X (-)

REFERENCES

1. Russ Em, Raymunt J. and Bars (1956) - J. Cl. Invest. 35, 133-144.
2. Seidel D., H.U. Buff, V. Fansek. V. Blegl. Chim. Acta 66 195-207 (1976)
3. Neto A.J.M. - Momento Mēdico 225, 13 (1977)
4. Seidel D., Alanpovic, P. Furman, Mc Conethy W. J. Cl. Investig., 49:2396 (1970)
5. Seidel D., Schmut E. A. and Alanpovic - Med. Wochenscher 15:671 (1970)
6. Jirgl V. - Kin Wschs 35:938 (1977)
7. Dias B., Tavares F., Serra A.V. - Coimbra Medica XV, 1085,1109 (1968)
8. Neto A.J.M. - Rev. Port. Bioquīmica Aplicada nº 2, 99 (1978)
9. Talafant E., Tavarek E. - Clin. Chim. Acta 96, 261-64
10. Burstein M., Samaille J.(1957) Rev. Hermat. 12, 678-80

ON THE DEVELOPMENT OF APOPROTEIN AND LIPID LEVELS THROUGHOUT CHILDHOOD

Walter F. Riesen, Rubino C. Mordasini,
Flavio D. Sepulcri, and Oskar H. Oetliker

Inst. Clin. Exp. Cancer Res. and Inst. Clin. Protein Res., Univ. Berne, Tiefenau-Hospital 3004 Berne, and the Div. Pediatric Nephrol.,Univ. Children's Hospital Berne, Switzerland

INTRODUCTION

Epidemiologic and genetic studies on the evolution of cardiovascular disease in man have repeatedly supported the importance of plasma cholesterol levels in the pathogenesis of coronary atherosclerosis (1-4). These studies also emphasize the importance of distinguishing the cholesterol content of the lipoprotein fractions, very low density (VLDL), low density (LDL) and high density lipoproteins (HDL). A close correlation between increased levels of LDL, known to contain the major part of the transported cholesterol, and the risk of developing coronary atherosclerosis was found (2,5). HDL, on the other hand, appeared to be negatively correlated with the occurrence of coronary heart disease (6-9).

In view of the importance of the two antipodes LDL and HDL for the pathogenesis of coronary atherosclerosis the development of these lipoprotein fractions throughout childhood is of particular interest.

The present study reports on the development of apoprotein and lipid levels throughout childhood.

Material and Methods

Patients: Blood samples consisted of cord blood of newborns, of venous samples from infants and children, as well as of venous blood from adult men.

With an exception in the infants at birth, the blood samples were taken after a 12 hours fast. None of the patients had any signs of an alteration in the metabolism of lipoproteins, proteins or glucose.

Methods: Cholesterol was measured in both serum and isolated lipoprotein fractions by the Liebermann-Burchard reaction as described by Levine (10) in a Technicon AA II analyser.

VLDL was separated from LDL and HDL by ultracentrifugation at density 1.006 g/ml. LDL was separated from HDL by precipitation with manganese chloride (0.05 M) and heparin (0.1%) (11).

Apo B was measured by radial immunodiffusion, using commercially available immunodiffusion plates (Behringwerke Marburg/Lahn, FRG) and the apo B standard from the same company.

Antibodies against apoproteins A-I and A-II (12) were quantitated by electroimmunodiffusion in agarose (1.5%) using diethylbarbiturate buffer, pH 8.2, $\mu = 0.1$ (13).

RESULTS

The levels of LDL and HDL cholesterol and the total cholesterol levels at birth, in different (arbitrarily chosen) age groups of children and in an adult collective of men are shown in Fig. 1.

The data show a drastic increase of total cholesterol and LDL cholesterol and a moderate increase of HDL cholesterol levels from birth to infants and children aged below 7 years. During puberty a net drop of total cholesterol and LdL cholesterol levels occurs. The mean HDL cholesterol level remains unchanged in this age group. A further important increase in total cholesterol and LDL cholesterol is noted in the adult collective. This increase appears to be only moderate in the HDL cholesterol levels.

In Fig.2 the levels of apoproteins B, A-I and A-II in the different age groups are given.

Apo B levels, similarly to LDL cholesterol levels, show an important increase after birth. Apo B levels show a tendency to decrease after 6 years of age and are markedly decreased during puberty. In adult men apo B levels are again increased, the mean level being similar to that observed in children aged between 3 months and 6 years. Apo A-I and A-II levels show a slight increase with age and thus largely reflect the concentrations of HDL cholesterol in the different age groups.

In Figs.3 to 6 the ratios of total serum cholesterol to LDL or HDL cholesterol or to the levels of the apoproteins B, A-I and A-II are plotted versus the age. While the ratios of HDL cholesterol to apo A-I and to apo A-II seem to remain rather constant throughout childhood into adulthood, the ratio of total cholesterol to apo A-I or A-II shows a considerable increase after birth. In contrast to the apoproteins A-I and A-II, the ratio of the LDL cholesterol to apo B shows an increase from childhood to the adult collective, indicating that the cholesterol binding sites of apo B are not fully saturated in childhood.

DISCUSSION

This report describes the development of total cholesterol levels, the distribution of cholesterol in low and high density lipoprotein fractions and the ratios between cholesterol and

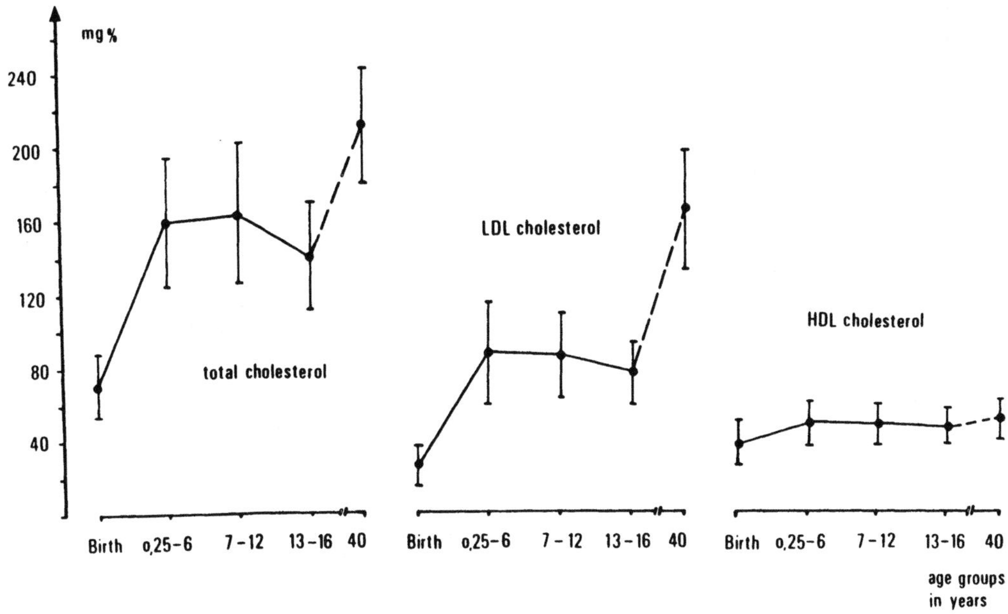

Figure 1. Total colesterol, LDL cholesterol and HDL cholesterol levels in cord blood and in sera from normal donors from 3 months to 40 years old.

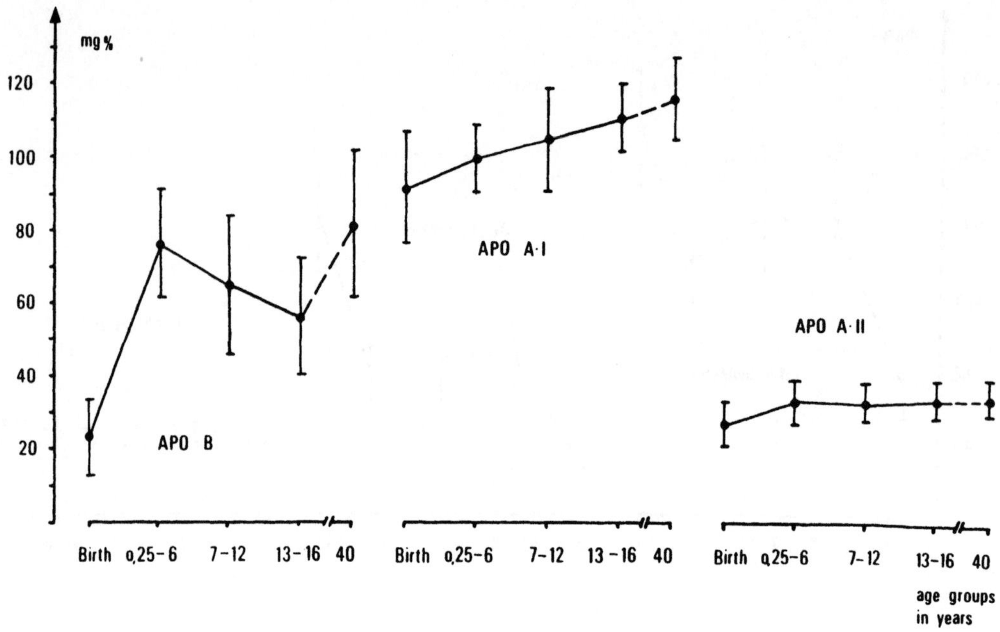

Figure 2. Apo B, A-I and A-II levels in cord blood and in sera from normal donors from 3 months to 40 years old.

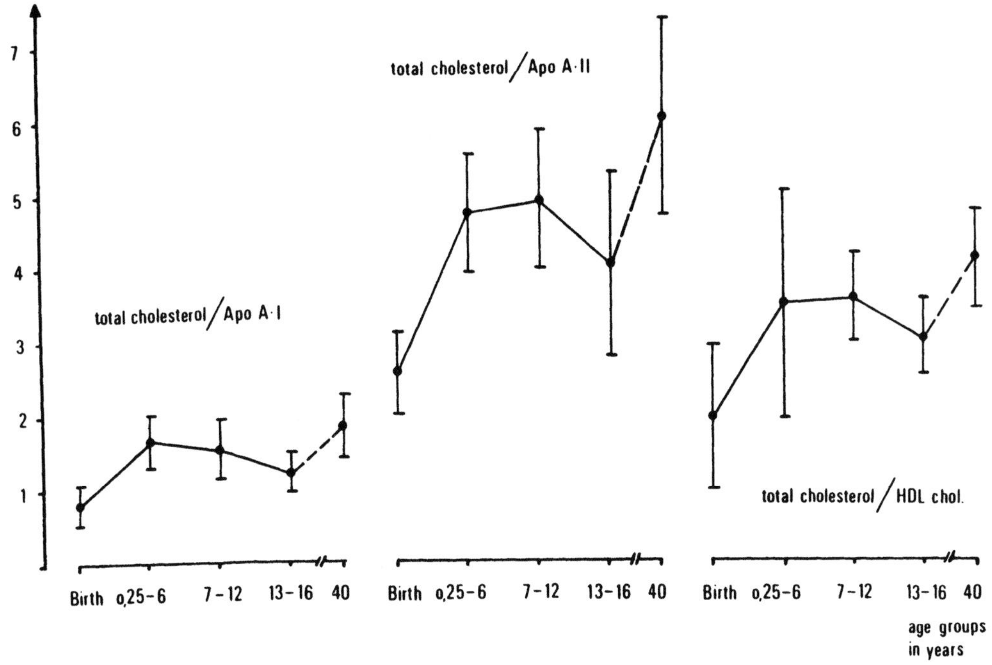

Figure 3. Ratios of total cholesterol to apo A-I, A-II and HDL cholesterol in cord blood and in sera from normal donors from 3 months to 40 years old.

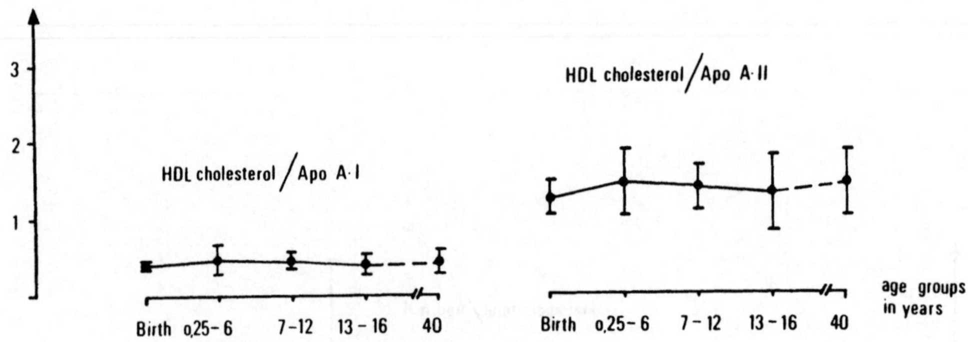

figure 4. Ratios of HDL cholesterol to apo A-I and A-II in cord blood and in sera from normal donors from 3 months to 40 years old.

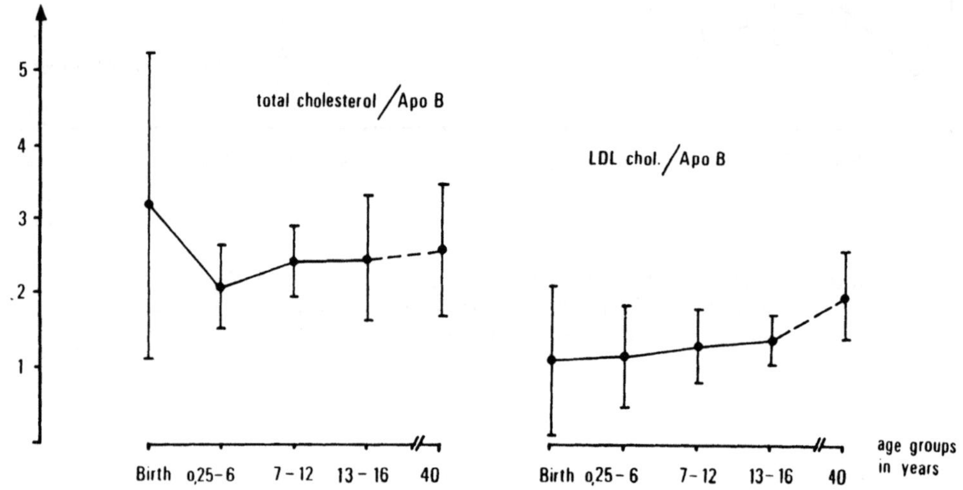

Figure 5. Ratios of total cholesterol and LDL cholesterol to apo B in cord blood and in sera from normal donors from 3 months to 40 years old.

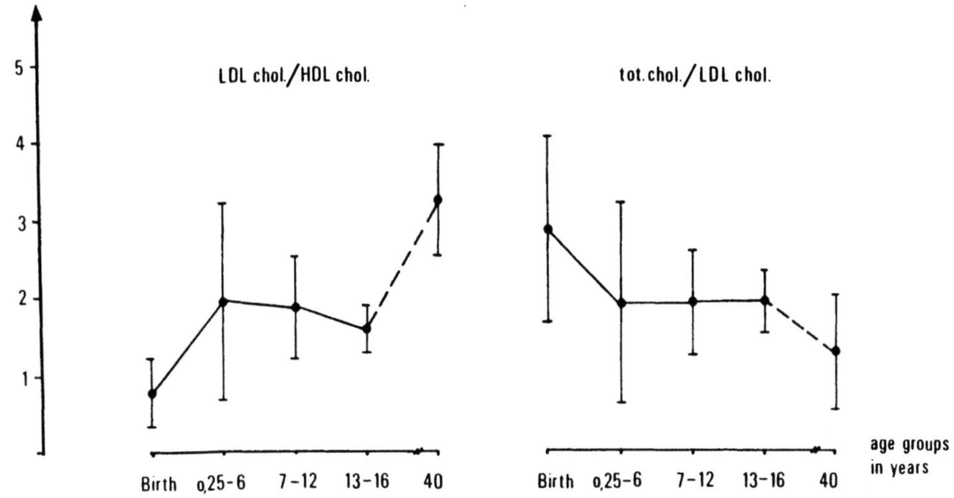

Figure 6. Ratios of LDL cholesterol to HDL cholesterol and of total cholesterol to LDL cholesterol in cord blood and in sera from normal donors from 3 months to 40 years old.

apoproteins A and B from birth throughout childhood and in an adult male collective.

Out data show that in cord blood, LDL cholesterol levels are only about one third of that observed in the serum of young children. This might indicate that LDL does not cross the placental barrier and that the measured LDL was synthesized by the fetus itself. At the age of three months to six years the peak values of childhood are found. After this age period, LDL cholesterol concentrations decrease, reaching lower levels during puberty, a finding which is consistent with recent data of Rifkind et al. (14). This period is followed by a doubling of the value in male adults. The course of the total serum cholesterol concentrations in the different age groups resembles that of the LDL cholesterol, but total cholesterol levels in cord blood are relatively higher than LDL-cholesterol levels, while in the adults they are relatively lower. This represents a repartition of cholesterol in favour of the atherogenic LDL fraction.

During childhood, apo B levels correspond to the LDL cholesterol levels. In adults, However, the mean concentration of apo B is only slightly above the highest value observed in children.

In contrast to LDL cholesterol, the concentration of the HDL cholesterol, which is known to be negatively associated with the risk of coronary heart disease (6-9), makes up for 80% of the total cholesterol observed at birth indicating that HDL may cross the placental barrier. HDL cholesterol remains rather constant throughout infancy and childhood, and shows but a slight increase in the adult male collective. In cord blood, the levels of two major apoproteins of HDL, apo A-I and apo A-II make up already for most of the total cholesterol concentration measured. Apo A-I and apo A-II gradually increase during childhood without showing a significant descrease during puberty as it is the case for apo B.

The most dramatic changes among all ratios under study occurs with the ratio of LDL to HDL cholesterol. While HDL cholesterol is prevalent in cord blood, there is a preponderance of LDL cholesterol in serum of children which increases to about three times more LDL than HDL cholesterol in adult men. Although total cholesterol values are considerably higher in adults than in children the ratio of total cholesterol to LDL cholesterol decreases from cord blood to sera of adults due to the enormous increase of LDL cholesterol.

Our data indicate that the most pronounced change in lipoprotein concentrations during ontogeny is the increase of LDL, which effects the ratio of LDL to HDL and the ratio of total cholesterol to apo A-I and A-II in disfavour of HDL, known to be negatively correlated with the risk of coronary heart disease. The mechanisms which underly this well-documented fact, however, are still under exploration. It may relate to the role of HDL in the transport of cholesterol out of cells or to the HDL mediated inhibition of LDL uptake and degradation by cells (15,16).

LDL is known to play a major role in the pathogenesis of atherosclerotic lesions. This noxious effect of LDL may be alleviated by the preexistence of high HDL levels or amplified by the low HDL cholesterol levels, which remains nearly unchanged during the first five decades of life. The importance of HDL within the development of atherosclerosis does not become evident from our study of a random collective of normal probands.

REFERENCES

1. McGee, D., and T. Gordon. April 1976. The Framingham Study: the results of the Framingham Study applied to four other U.S.-based epidemiologic studies of cardiovascular disease. Section 31, DHEW Public No. (NIH) 76-1083.
2. Kannel, W.,B. W.P. Castelli, T. Gordon, and P. McNamara. 1971. Serum cholesterol, lipoproteins, and the risk of coronary heart disease. The Framingham Study. Ann. Intern. Med. 74: 1-12.
3. Keys, A. 1975. Coronary heart disease - the global picture. Atherosclerosis 22: 149-192.
4. Gordon, T., M. Carcia-Palmieri, A. Kagan, W.B. Kannel, and Scherffman, J. 1974. Differences in coronary heart disease in Framingham, Honolulu and Puerto Rico. J. Chronic. Dis. 27: 329-344.
5. Walton, K.W. 1975. Pathogenic mechanism in atherosclerosis. Amer. J. Cardiol. 35: 542-558.
6. Castelli, W.P., J.T. Doyle, T. Gordon, C.G. Hames, M.C. Hjortland, S.B. Hulley. A. Kagan, and W.J. Zukel, 1977. HDL-cholesterol and other lipids in coronary heart disease. Circulation 55: 767-772.
7. Gordon, T., W.P. Castelli, M.C. Hjortland, W.B. Kannel, and T. Dawber. 1977. High density lipoprotein as a protective factor against coronary heart disease. The Framingham Study. Amer. J. Med. 62: 707-714.
8. Miller, N.E., O.H. Forde, D.S. Thelle, and O.D. Mjos. 1977. The Tromsø heart-study- High-density lipoprotein and coronary heart-disease. A prospective case-control study. Lancet i: 965-967,
9. Salel, A.F., A. Fong, R. Zelis, R.R. Miller, N.O. Borhani, and D.T. Mason. 1977. Accuracy of numerical coronary profile. Correlation of risk factors with arteriographically documented severity of atherosclerosis. New Engl. J. Med. 296: 1447-1450.
10. Levine, J., S. Morgenstern, and D. Vlastelica. 1967. Direct Liebermann-Burchard method for serum cholesterol. Technicon Symp. 1: 25.
11. Burstein, M., and H.R. Scholnik. 1971. Isolation of lipoproteins from human serum by precipitation with polyanions and divalent cations. Prot. Biol. Fluids 19: 21-27.
12. Riesen, W.F., R.C. Mordasini, and G.W. Middelhoff. 1978. Quantitation of the two major apoproteins of human high density lipoproteins by solid phase radioimmunoassays. FEBS letters 91: 35-39.

13. Mordasini, R., and W.F. Riesen. 1980. Electroimmunoassay and radioimmunoassay for the quantitation of high density apolipoproteins A-I and A-II. J. Clin. Chem. Clin. Biochem. J. Clin. Chem. Clin. Biochem. 18: 917-920.
14. Rifkind, B.M., I. Tamir, G. Heiss, R.B. Wallace, and H.A. Tyroler. 1979. Distribution of high density and other lipoproteins in selected LRC prevalence study populations: a brief survey. Lipids 14: 105-112.
15. Stein, Y., M.C. Glangeau, M. Fainaru, and O. Stein. 1975. The removal of cholesterol from aortic smooth muscle cells in culture and Landschutz ascites cells by fractions of human high density apolipoprotein. Biophys. Acta 380: 106-118.
16. Carew, T.E., S.B. Hayes, F. Koshinsky, P. Steinberg. 1976. A mechanism by which high density lipoproteins may slow the atherogenic process. Lancet: 1315-1317.

Acknowledgments:

The authors thank Miss E. Ischi, E. Schlumpf and I. Schmid for expert technical assistance. This work was supported by the Swiss National Science Foundation.

LIPOPROTEINS AS RISK FACTORS

Howard A. Eder

Professor of Medicine
Albert Einstein College of Medicine

New York, New York

INTRODUCTION

The Relationship Between the Different Lipoprotein Classes and Atherosclerosis

This paper reviews the role of the lipoproteins as risk factors, the metabolic relationship between them and how this contributes to the pathogenesis of atherosclerosis. The conclusion to be drawn is that lipoproteins should not only be considered individually as risk factors but also as part of a dynamic system whereby metabolic products of lipoproteins may be atherogenic.

Evidence that atherosclerosis may be reversible in man is encouraging to scientists working in this field. There are data from a variety of sources indicating that atherosclerosis may be a reversible process. Such data include evidence from European countries that were blockaded during the second world war, as well as differing disease incidences among different populations.

Recent data from the US show that the incidence of cardiovascular disease has been falling since the late 1950s and this decrease has increased since 1967. All cardiovascular disease is shown in the figure but the major contributor to this decline in mortality has been the fall in incidence of coronary heart disease. We would all like to claim that this has been due to some aspect of our work, but that claim cannot really be justified. However, part of the basis for such a claim is the shifting level of serum cholesterol in the U.S., shown in three seperate surveys. These show that there has been a distinct decrease in the serum cholesterol level in the large populations sampled, (e.g. the LRC population is a sample obtained after screening 60,000 people). These data suggest that changes in serum cholesterol may play a role in the decreasing mortality from cardiovascular disease, but the reason for the decrease is unknown.

The fact that death from cardiovascular disease is decreasing suggests that the process is reversible and that research directed to the problem of atherogenesis is very relevant and that measures taken to reduce risk factors in a population might be useful. Certainly levels of plasma lipoproteins do effect risk from C.H.D.

Plasma Lipoproteins

The very largest particles (diam. 1,000-10,000 Å) the chylomicrons, enter the circulation in normal subjects following the ingestion of a fatty meal.

These particles consist largely of triglyceride with small amounts of protein, cholesterol and phospholipid. In certain diseases they are present in the fasting state.

The VLDL (250-750 Å) are present in fasting man and most of the triglyceride in healthy fasting individuals are transported in this fraction. They have large molecular weight (10^7) Daltons but not as large as chylomicrons (10^8) Daltons.

The IDL (250 Å:MW, 4.5 $X10^6$) are particles which are not normally present in fasting man and have a transitory existence in the circulation following the ingestion of a fatty meal. In one disease, (Type III hyperlipoproteinaemia) they do occur.

LDL (220-230Å) is a particle with a molecular weight of $2.3X10^6$ which transports about two thirds of the plasma cholesterol and is certainly the most malevolent of the various lipoprotein particles in the pathogenesis of atherosclerosis.

HDL may be separated into sub-fractions: HDL_2 which has a molecular weight of about 380,000 and HDL_3which has a molecular weight of 180,000. The HDL_2 is the larger particle containing relatively more lipid than protein (40%) whereas HDL has around 55% protein.

Apo-Proteins

Most lipoproteins contain a mixture of heterogeneous proteins; the apolipoproteins Apo-AI and Apo-AII are the major proteins of HDL.A-I is synthesized in the intestines and probably also in the liver: it activates the enzyme lecithin cholesterol-acyltransferase (LCAT). Apo-B has been shown to be heterogenous by SDS gel electrophoresis. It consists of two major components, a higher molecular weight protein found largely in the liver and a lower molecular weight component found largely in the intestine. The molecular weights of these components are more than 250,000. Apo-B is the major apoprotein of VLDL and the sole protein of LDL. The C-apoproteins are of low molecular weight (6,000-9,000) and the one of major interest is the C-II which activates the enzyme lipoprotein lipase. This enzyme hydrolyses triglyceride and is responsible for the initial step in the catabolism of chylomicrons in VLDL. Recently families with a congenital absence of C-II have been identified and these patients are very similar to patients who have lipoprotein lipase deficiency (Type I). Apo-D is found in HDL and is involved in the LCAT reaction in a quite complex way. Apo-E is involved in cholesterol transport and there are specific tissue receptors that recognize both Apo-E and also Apo-B. Apo-E exists in multiple isoforms. In Type III hyperlipoproteinemia certain of the isoforms which are recognized by hepatic receptors are absent leading to the accumulation of B-VLDL in the plasma.

Metabolism of Lipoproteins

When lipid is absorbed by the intestine, chylomicrons containing triglyceride are synthesized by the intestinal epithelial cells and secreted into the intestinal lymph. They contain chiefly triglyceride, Apo-B and also some of the proteins which are usually associated with HDL, namely, A-I, A-II and A-IV. When the chylomicrons enter the circulation the A-I, A-II, and A-IV rapidly leave the chylomicrons which acquire Apo-E and C-apoproteins which are synthesized in the liver. The chylomicrons is attacked by the enzyme, lipoprotein lipase, much of the triglyceride is hydrolysed and the particle becomes smaller. As it does so, some of the surface components such as various apoproteins and phospholipids are removed and some of these constituents can be recovered in HDL. The particle remaining, the chylomicrons remnant, is taken up by the liver which contains surface receptors for the Apo-B and Apo E present in the remnants. In man the VLDL remnant is converted to LDL. Zilversmit (1) has suggested that the chylomicron remnants may be an atherogenic particle. The finding that cells of the arterial wall have high affinity receptors for Apo-B and Apo-E support this hypothesis. Recently it has been shown that macrophages have high affinity receptors for Apo-E suggesting that the foam cells found in atherosclerotic plaques may be formed by this mechanism.

Although VLDL can be made by the intestines, in man probably most of it is made by the liver. In healthy subjects the Apo-B of VLDL is the source of the Apo-B in LDL. Thus VLDL is atherogenic in the sense that it is the precursor of LDL.

In recent years a great deal has been learned about the metabolism of LDL, through the classic work of Goldstein and Brown (2), in studying the cellular uptake of LDL, used as controls, in patients with homozygous familial hypercholesterolaemia. They showed that normal cells could bind LDL when they were cultured in a media that was depleted of LDL, whereas cells from patients with homozygous familial hypercholesterolaemia could not bind LDL. They developed the concept of the LDL pathway, which is that certain cells, including fibrobasts, lymphocytes, adrenal cells, ovarian cells and others, have the capacity to bind LDL and internalize it. When it is internalized, the protein is degraded and the cholesteryl ester hydrolysed. In certain tissues, such as the adrenal or ovary, this cholesterol is the precursor for steroid hormones. In other tissues, the cholesterol is stored and this regulates the cellular content of cholesterol both by down regulating HMG CoA reductase and the availability of LDL receptors. When the cell is not exposed to LDL it makes additional receptors so that when LDL is present it is bound and internalized. One of the aspects of this system of regulation must be the removal of cholesterol from cells. As there are systems for accumulating cholesterol in a cell, there must also be systems for cholesterol to leave it. It was first suggested by Glomset (3) that HDL plays a role in transferring cholesterol from cells but this hypothesis still lacks adequate proof. Furthermore the net result of this may not lead to a reduction in cellular cholesterol since recent studies have suggested that the cholesterol transferred to HDL is esterified and then translocated to VLDL which is metabolised to LDL which is taken up by cells. The role of HDL in removing cholesterol from cells is one that requires further research.

Risk Factors in CHD

1. LDL: Many studies have shown that elevated serum cholesterol is a major risk factor. In most instances the level of LDL correlates well with the level of serum cholesterol. The exception is those patients with very high levels of HDL. Thus the data on risk due to the level of serum cholesterol actually indicates risk due to the concentration of LDL. The Framingham data clearly demonstrates the importance of LDL as a risk factor.

2. VLDL: There is a great deal of disagreement about the importance of VLDL as a risk factor. Since the concentration of serum triglyceride is highly correlated with the level of VLDL, both measurements can be interchanged similarly to LDL and serum cholesterol. Since epidemiologists are concerned with defining primary risk factors they tend to dismiss VLDL as a risk factor since some studies using multivariate analysis can demonstrate that the risk from elevated VLDL is secondary. On the other hand most studies show that elevated VLDL is a risk factor, either primary or secondary. There are data from prospective studies showing elevated VLDL (or triglyceride) to be a risk factor and studies of survivors of myocardial infarction have demonstrated a higher prevalence of hypertriglyceridaemia in such patients. A physician has to consider the patient as an individual and does not perform multivariate analysis. When VLDL or serum triglyceride is elevated in a fasting subject, the physician should consider them at increased risk and undertake measures that will decrease VLDL and triglyceride.

3. HDL: In 1951 we compared HDL levels in healthy subjects with levels in survivors of myocardial infarction (Barr, *et al.* 4) and found that the latter group had lower levels of HDL than did normal subjects. In 1975 Miller and Miller (5) suggested that HDL might be a protective factor against coronary heart disease. Subsequently a large number of studies of this relationship were performed. The Cooperative Phenotyping Study (6) studied populations in five cities and found that in the individuals with lower levels of HDL the prevalence of CHD was considerably higher than in individuals with higher levels of LDL. This was true up to HDL cholesterol levels of 45-54 mg/dl and there was no further gradient with higher concentrations. The Framingham project (7) began measuring HDL twenty years after the study was started so that those subjects were older, ran-

ging in age from 49 to 82. This study had the advantage of being a prospective study. The incidence of CHD increased steadily as the HDL level fell both for men and for women. From these data, it would be concluded that, for a drop in HDL, from a normal value in a male of 45 mg cholesterol/100 ml at 35, there is a 50% increase in risk. However, it must be emphasized that this study was carried out in an older population so that the findings cannot necessarily be extrapolated to younger populations. Subsequent studies have been performed in younger populations but the number of events was small.

Studies of other complications of atherosclerosis such as cerebrovascular disease and peripheral vascular disease had not shown any consistent correlations with serum lipids and lipoproteins. It is of considerable interest that patients with cerebral vascular disease had lower HDL than did controls. Recent studies have also shown that patients with peripheral vascular disease have low HDL levels, but this is complicated by the frequency of diabetes in these patients.

There are many factors which effect the level of HDL (Table 1). Some of the important factors that lower HDL are oral contraceptives, diabetes, smoking, certain anti-hypertensive drugs, certain diets and Tangier's disease. Populations with a low fat intake and low levels of serum cholesterol (e.g. Mexican Indians who have a very low incidence of CHD) have low levels of LDL, but also very low levels of HDL. HDL-cholesterol was also found to be low among vegetarians in the USA who have lower levels of HDL than non-vegetarians. Thus the absolute level of HDL is probably not critical, but if HDL does play a protective role it is probably in relation to the levels of LDL.

HDL is heterogenous and can be separated into a number of different subclasses. Initially HDL was separated into two subclasses, HDL_2 and HDL_3. HDL_2.is larger, has a lower density and contains more lipid than HDL_3. It is time consuming and expensive to perform these separations by conventional methods. Recently we have developed a method in our laboratory (8) in which HDL_2 is precipitated by dextran sulfate in the presence of Mn^{++}. This method is simple, inexpensive, and rapid. In Table 2 are shown data which we obtained. In a healthy population of medical students and people working in the medical school women had higher levels of HDL as is well known; HDL_2 was also higher in women than in men. When HDL varied the change was largely but not entirely in HDL_2. This can be seen in a group of runners who had total serum HDL averaging 60 mg/dl as opposed to controls (46 mg/dl), the HDL_2 was doubled, whereas the HDL_3 showed little change. Similarly, in subjects eating diets with a high P/S ratio total LDL but also HDL decreased. The HDL_2 fraction dropped to a greater extent.

Measurement of HDL_2 may therefore be a more sensitive indicator of change than measurement of total HDL. Whether or not it turns out to be of practical value we are not certain at the present time. However it has been demonstrated to be useful in studying factors affecting total HDL.

Table 1 Some Factors Influencing HDL Levels

Increase	Decrease
Women	Men
Russians	American
Black Americans	White Americans
Exercise	CHD
Oestrogens	Androgens
Insulin	Progestagen (OCs)
Alcohol	Diabetes
Clofibrate	Oral Hypoglycaemics
Nicotinic acid	Smoking
Chlorinated hydrocarbons	Anti-hypertensives
Phenytoin	Obesity
Weight Reduction	Chronic Uraemia
Familial Hyper Alpha	Polysaturated fat diets
	High carbohydrate diet
	Hypertriglyceridaemia
	Tangier Disease

Table 2 HDL and Sub Fractions in Normal Healthy Men and Women

Men (No)	HDL	HDL_2	HDL_3
	(Mg cholesterol/dl)		
Age 20 - 34 (142)	44 ± 10	15 ± 7	28 ± 7
35 - 49 (89)	44 ± 13	12 ± 8	30 ± 7
50 - 64 (65)	46 ± 14	15 ± 10	30 ± 12
Women (No)			
Age 20 - 34 (60)	52 ± 12	20 ± 11	31 ± 5
35 - 49 (49)	52 ± 12	17 ± 11	34 ± 5
50 - 64 (61)	57 ± 17	20 ± 12	35 ± 9

Table 3 HDL and Sub-classes in 37 Men with CHD (mean ± 50)

	mg cholesterol/dl		
	HDL	HDL_2	HDL_3
Healthy men (50-64)	46 ± 14	15 ± 10	30 ± 12
CHD (40-69)	32 ± 10	6 ± 6	25 ± 5

Table 4 Effects of Treatment on Plasma Lipids and HDL in NIDDM

		Age	TC	TG	HDL-C	HDL_2-C	HDL_3-C
Insulin	5 years						
	Men	52	209	108	40	11	31
	Women	51	231	232	37	10	27
Insulin	5 yrs						
	Men	51	236	119	52	23	30
	Women	555	209	239	46	16	30
Sulphonylureas							
	Men	58	225	110	38	9	29
	Women	57	249	213	40	10	30

Clinical disorders with Low HDL

In a group of ambulatory patients who underwent coronary angiography and were found to have changes in their coronary arteries, the HDL was low and the HDL_2 was even more markedly reduced, but there was also a slight reduction in HDL_3 (Table 3). Patients with insulin-dependent diabetes have reduced HDL and markedly reduced levels of HDL_2. Non-insulin-dependent diabetics also have low HDL and the interesting thing is that women have HDL and HDL_2 values which are very comparable to those in men. As women usually have higher HDL and HDL_2 levels than men, this means that the reduction in women was greater than in men, so that the increased incidence of CHD in women, or the greater susceptibility of women with diabetes, to CHD may possibly be related to this difference in HDL. We are now accumulating data which suggest that even with a few months intensive insulin therapy, HDL_2 may be increased and in many patients followed for over five years HDL is almost normal. HDL_2 is normal with very little effect on HDL_3. Table 4 shows data from a group of patients with insulin-dependent diabetes, and the effect of insulin and sulfonyl urea of HDL. The data on sulfonyl urea administration are of interest because there have been a number of studies suggesting that these drugs lower HDL. This would be very significant in relation to the UGDP Study which concluded that oral hypoglycaemic agents increased the risk of cardiovascular disease. Studies which we have carried out to date give no suggestion that oral hypoglycaemic agents neither lower or raise HDL or HDL_2 in diabetics whereas insulin administration usually increases HDL and HDL_2.

HDL might be protective inthat it removes cholesterol from tissues or it might be protective because it competes with LDL for uptake by tissues. A third hypothesis is that HDL is synthesized as a nascent particle, both in the liver and intestine. These nascent particles then receive surface components from the metabolism of chylomicrons which lead to increases in HDL_2 concentrations. Thus in a sense HDL is a reflection of the efficacy of the system for handling ingested fats. When HDL is present in higher concentrations the disposal of ingested fat is efficent whereas a low HDL_2 concentration may indicate a less efficient system for chylomicron removal.

SUMMARY

Recent studies of the metabolism of lipoproteins and of mechanisms of atherogenesis have led Steinberg to propose a unified concept of atherogenesis. Increased levels of LDL cause infiltration of lipoproteins into tissue with cholesteryl ester accumulation. Endothelial injury leads to platelet aggregation, production of platelet growth factor, smooth muscle proliferation and formation of lesions. There is an interplay between these factors so with LDL causing endothelial injury and with endothelial injury there can be increased infiltration of LDL into tissue. LDL can cause platelet aggregation which results in proliferation of smooth muscle. LDL can also cause proliferation of arterial smooth muscle. There is accumulation of macrophages and some of these macrophages have affinity for Apo-E containing proteins, suggesting that chylomicrons remnants may play a role. HDL may play a role in removing cholesterol from tissue, but the major atherogenic lipoprotein is LDL and certainly therapy should be directed primarily at reducing levels of LDL.

REFERENCES

1. Zilversmit, D.B. Circulation 60:473-485, 1979
2. Goldstein, J.L. and Brown, M.S. Ann. Rev. of Biochem. 46:897-930, 1977
3. Glomset, J.A. J. Lipid Res. 9:115-167, 1968
4. Barr, D.P., Russ, E.M., and Eder, H.A. Amer. J. Med. 11:480-493, 1951
5. Miller, G.J. and Miller, N.E. Lancet i, 16-19, 1971
6. Castelli, W.P., Doyle, J.T., Gordon, T., James, C.G., Hjortland, M.C., Hulley, S.B., Kagan, A., and Zukel, W.J. Circulation 5: 767-772, 1977
7. Gordon, T., Castelli, W.P., Hjortland, M.C., Kannel, W.B. and Dawber, T.R. Amer. J. Med. 62: 707-714, 1977
8. Gidez, L.I., Miller, G.J., Burstein, M. and Eder, H.A. In Report of the HDL Methodology Workshop N.I.H. Publication Nº. 79:1661, 1979.

INDEX

GPSR Compliance

The European Union's (EU) General Product Safety Regulation (GPSR) is a set of rules that requires consumer products to be safe and our obligations to ensure this.

If you have any concerns about our products, you can contact us on ProductSafety@springernature.com

In case Publisher is established outside the EU, the EU authorized representative is:

Springer Nature Customer Service Center GmbH
Europaplatz 3
69115 Heidelberg, Germany

Batch number: 10684222

Printed by Printforce, the Netherlands